It’s Always about Leadership

IT'S ALWAYS ABOUT LEADERSHIP

DENNIS L. RUBIN

Disclaimer

The recommendations, advice, descriptions, and methods in this book are presented solely for educational purposes. The author and publisher assume no liability whatsoever for any loss or damage that results from the use of any of the material in this book. Use of the material in this book is solely at the risk of the user.

PennWell Corporation
1421 South Sheridan Road
Tulsa, Oklahoma 74112-6600 USA
800.752.9764
+1.918.831.9421

sales@pennwell.com
www.FireEngineeringBooks.com
www.pennwellbooks.com
www.pennwell.com

Publisher: Matthew Dresher
Managing Editor: Mark Haugh
Production Manager: Sheila Brock
Production Editor: Tony Quinn
Cover Designer: Beth Rose
Book Designer: Susan E. Ormston

Library of Congress Cataloging-in-Publication Data

Names: Rubin, Dennis, author.
Title: It's always about leadership! / Dennis L. Rubin.
Description: Second edition. | Tulsa, Oklahoma, USA : PennWell Corporation, 2018. | Includes bibliographical references.
Identifiers: LCCN 2018012930
Subjects: LCSH: Fire departments--Management. | Fire departments--Employees--Coaching of. | Leadership.
Classification: LCC TH9158 .R83 2018 | DDC 363.37/80684--dc23

Printed in the United States of America

1 2 3 4 5 22 21 20 19 18

It is not the critic who counts; not the man who points out how the strong man stumbles, or where the doer of deeds could have done them better. The credit belongs to the man who is actually in the arena, whose face is marred by dust and sweat and blood; who strives valiantly; who errs, who comes short again and again, because there is no effort without error and shortcoming; but who actually strives to do the deeds; who knows great enthusiasms, the great devotions; who spends himself in a worthy cause; who at the best knows in the end the triumph of high achievement, and who at the worst, if he fails, at least fails while daring greatly, so that his place shall never be with those cold and timid souls who neither know victory nor defeat.

—Theodore Roosevelt, 26th President of the United States
Excerpt from the speech "Citizenship in a Republic,"
December 8, 1905

This book is dedicated to all firefighters—past, present, and future.

CONTENTS

FOREWORD

It has been my distinct privilege to count Chief Dennis L. Rubin as a friend and colleague for over 30 years. During that time, we have shared problems, solutions, ideas (some good, some not so good) as we have moved through our individual fire service careers. Most of that time, we worked on opposite sides of the country in very different fire departments. While we both worked in the Metro Phoenix area for a period of time in the 1980s, even that was in vastly different fire departments.

Fig. F–1. Bruce H. Varner, fire chief

My principal reason for pointing this out is that the rules and solutions that Chief Rubin is offering in this book are among those that we have shared, and the differences in our fire departments, their sizes or geographic locations, really didn't matter. What matters is that Chief Rubin has been moving around the fire service (just a bit) and has had the opportunity to gain those experiences in a variety of fire departments, ranging from combination and fully paid to large metropolitan stations (including those in our nation's capital, where both of us were born and which every day brings new, interesting, and sometimes very exciting challenges).

The principles offered in this book are sound personal and professional advice; whether you are a newbie or an experienced old salt, there are lessons to take away with you. I hope you enjoy reading the book as much as I did and find the offerings worthy of use in your life and organization.

—Bruce H. Varner, fire chief

INTRODUCTION TO THE SECOND EDITION

Over several decades in the fire service business, I have developed a list of rules that I try to live by on and off duty. I kept this list of ideas on three-by-five cards and reviewed the information on them from time to time. Typically, I would pull out the cards when I was preparing for promotional interviews or simply wanted to make sure that I was focusing on what was important. The cards served me well, and in great anonymity until recently.

A fire-rescue service mentor asked me to make a presentation on leadership. The specific request was to present my leadership style and philosophies to a national group of senior fire and EMS officers at a staff and command school. That is when I took out the cards and developed a successful presentation based on the rules that I had jotted down over the years. Those who attended the first session of "Rube's Rules: A Leadership Journey" enjoyed the information and provided positive feedback. The rules listed here are in no special order, and all are of equal value.

Since then, I have presented this material many times, refining the content after each delivery. I thought that it would be valuable to our profession to take the next logical step by documenting these rules for publication so that more folks could have the benefit of this experience. By no means are these 13 rules new or breakthrough information; they are best described as time-tested leadership points with my personal twist. If you implement them in your daily leadership and management style, I am certain that you will improve your performance as a courageous leader within your organization.

The first edition of this book was titled *Rube's Rules for Leadership*. It represented a great collection of the rules that I think are mission-critical to running a fire station or a fire department or simply managing yourself. These 13 important rules are best known as "Rube's Rules." I have put this information together over the past 30 or so years of my fire-rescue career. Based on the popularity of this book and the associated presentations, I have expanded the material in this textbook to include four new subsections in each chapter: (1) Critical Learning Points, (2) Leadership Discussion Questions, (3) Personal Leadership Plan for Improvement, and (4) Practical Application and Related Case Studies. The idea is that the reader (learner) can put this book into action to improve their leaderships skills at all levels at any time in any field.

I have worked diligently to adopt each and every rule to be a better person, firefighter, and leader of women and men. I would say that there is nothing earth-shattering about this list of rules that I offer. The significance of this work is my interpretation and application of these rules to everyday life (at the firehouse, home, or anywhere). It is my hope that these 13 rules will keep readers out of all types of trouble, at and away from their firehouse. Not a day goes by that we don't hear about members of our profession who have done something stupid, corrupt, or illegal and is in very serious trouble because of their foolish actions. I wish I could say it is only the newest of our firefighters who find themselves in difficult circumstances, but it is a wide range from young to old, low-ranking to the chief, and just about every member in between. Hence the need to review the information in this book regularly.

You will also find, at the end of each chapter, critical learning points that were covered in the reading and leadership review questions. The vision is that after you complete this book, you will be able to easily and frequently return to the critical learning points for a quick refresher. The road to true leadership is and will always be about taking care of people (the customers and your workers). The need to renew that commitment and always be on top of your game is self-evident. The formal or informal boss needs to put in the work to complete that aspect of being a good leader.

The leadership review questions will be for deeper study and understanding of the information contained within a chapter. The hope is that this book will be used as the text for fire-rescue officer college courses and for the reader to continue the study of leadership traits and techniques. Further, the questions in this section will make for great firehouse dinner table discussion for evening and indoor drills. It is my desire that your copy of *It's Always about Leadership!* be dog-eared, highlighted, and very well worn. Heck, what is more enjoyable to a leader than preparing people for more responsibility in their respective organizations? This book should help that process work easily and effectively. And, by the way, you are all my people! (This concept is egregiously stolen from one of my favorite TV personalities, Al Roker of *NBC Today*. And yes, I am one of Al's people!)

These rules just may keep you and your department out of hot (and in some cases boiling) water. If a member adopts these leadership rules as their own, reviews them often, and lives by them, I think that the reader will enjoy a wonderful career without many issues or much drama. I can tell you that when a member "acts out" (does something wrong—immoral, illegal, or against fire service policy), every member in that department suffers. I understand that the punishment inflicted is transmitted primarily to the person who acts out, but everyone comes out a loser at some level. In fact, I would even go so far as to say that the entire fire-rescue service takes a hit when one of us does something stupid, immoral, illegal, or against policy.

Fig. I–1. This plaque is from the quarters of Engine 10 and Truck 13 in Washington, DC. I was assigned there in the 1970s.

The powerful collection of rules in this book will help everyone stay on the right course and make their departments proud to call them members. There are several ways that you can use the information offered in this textbook. First, it is simply just a good read, so take the time to enjoy it for what it is: a book written by a firefighter for firefighters. Next, it will serve as an ever-present reminder to go back and review the information inside the covers to stay on top of your A game. Finally, for all those who provide leadership (or are preparing to provide leadership), it is a great format to teach from and base discussion on with a company, a battalion, a division, or the entire department.

A suggestion is that the company officer should review one rule per shift at the dinner table, in an informal and comfortable setting. The company boss could conduct a detailed guided discussion based on the "rule reading" and add personal experiences to enrich the discussion while allowing the other members to add their experiences as well. As a long-time fire-rescue practitioner, if I can help a single firefighter avoid a pitfall (minor or major) that will get them "jammed up" (disciplined by superior officers or other authorities), I will do so.

From every angle, preventing dumb stuff from happening in the first place is just as important as preventing injuries at an emergency incident. This may sound a little overstated, but I firmly believe that it is time to work on our on-duty and off-duty behavior and performance. Just ask anyone who has been in this kind of trouble (maybe fired or removed from the rolls of the department as a volunteer member). I am sure that is a horrible feeling and should be avoided in the first place, if possible. Prior interventions to avoid inappropriate personnel actions would be a thousand times better than being on the outside looking in.

THE PROCESS

The reason I decided on 13 rules is fairly simple. I was told that I need to end with 12 or 14 rules, but not 13. I am guessing this was because of someone's personal superstition. However, after thinking quite a bit about this feedback, I could not cut a single rule, nor could I determine a need to add another rule, so 13 shall continue to be the number of leadership rules addressed in this book. I asked a very close and well-versed friend to take a look at the overall structure of this book, and he agreed with me that 13 seems to work perfectly well. In fact, this fire service icon's first reaction was that since you only need to follow 10 rules to get to heaven, following 13 rules should be just the right amount to make someone a good leader. I really liked his review and response, so the book remained as written.

I have been questioned about the importance and prominence of each of the 13 leadership rules. The fact of the matter is, each one is just as important as the others. I would place all in the *must follow and use* category. The reason they are in their current positions in this book is because that is how I wrote each one down as I refined the information. Do not overthink this book's contents. Each rule is written in a very understandable and direct fashion to be applied in the same way. Use all 13 rules and do not pick and choose the rules that are appealing to you.

I hope that each reader can attend a live leadership presentation of this book. The discussions are very rich in content, and the classroom lecture places all of the rules into context with the overall concept of leading. The class feedback and input are always amazing and the reason why this book is being released as an expanded edition (great few years for the first run).

You might be asking, Why a second edition of this simple textbook? After delivering dozens of training presentations based on the information captured in these pages, two requests kept surfacing. First, that I add more examples and exercises that would be useful for the reader. Second, that I make this book more connected to folks in other businesses and walks of life. I made two presentations for the City of Boca Raton Fire Rescue Service. Their fire chief, Tom Wood, asked the city's director of public works to sit in and be a course participant. That proved to be very successful for all involved. The director was insistent that I should broaden the scope of this work and make it more inclusive for all involved in both the public and private sectors. So, here goes the revised version. In an effort to avoid confusion, I have changed the title to *It's Always about Leadership!*, a phrase I use repeatedly. A tip of the fire helmet and a huge thank-you to Chief Tom Wood (fig. I–2) of Boca Raton and all of his amazing staff members!

Fig. I–2. Chief Tom Wood (left), Boca Raton Fire Rescue Service, with the author.

Safe firefighting and good government to all and may the god of your beliefs watch over you and your family at all times!

—Dennis L. "The Rube" Rubin

1

YOU HAVE TO SHOW UP

"Whatever you do, do it well."

—*Walt Disney*

The first rule seems simple enough, but never underestimate its value or power. Many functions, responses, and events occur in and around your fire department. By showing up to them, you raise your profile, your image, and your value within your organization. The phrase "showing up" has taken on yet an even more important meaning to organizational leadership: being present and available and showing you care about the members and the organization.

In 2011, a fire department out west was unable to assist an attempted suicide because of the circumstances of the request for assistance. The situation began with a mentally unstable person who determined that drowning himself was an acceptable way to end his life. The fire department was called to rescue that person. However, when fire service members arrived on the scene, they found the individual shouting out for someone to pluck him out of the ocean and return him to safety. The department was not able to assist because it lacked water rescue training and equipment, and so emergency responders were watching from the shoreline, as reported by some bystanders. Eventually a civilian good Samaritan pulled the man from the water. At this point, the fire department was able to provide medical care and transportation to a healthcare facility (the victim later died). In the view of the community, the fire department did not show up when needed during this situation. ("Alameda Fire Crews Watched Man Drown for Nearly One Hour, Transcripts Show," Bay City News Service, June 8, 2011.)

In a more recent case, a citizen caught the attention of an on-duty member in the front of a fire station. The verbal request (screaming) was made, asking for assistance with a cardiac case—a heart attack—in progress. The man was unconscious and unresponsive in the parking lot across the street from this fire station. The fact of the matter is that a fire truck and ambulance had to respond from a distance to help, while the ladder truck right across the street, staffed by five people, never left quarters. The outcome was the loss of human life: the 77-year-old man was pronounced dead upon arrival at a local hospital. The reason the fire company did not respond is unclear but there appears to have been a lack of leadership. One must ask what the leaders were thinking.

Leadership was clearly absent in both case studies. Let me start by pointing out that no one in a leadership role should be a micromanager unless the folks being led are new or in need of training or personal development. Additional information about that will come under several other rules. Your presence should send a loud and clear message that the leader is interested in the workers and the work. The followers want to know you care and see them as important. One of the many positive effects that this process has is to send this focused message to all: to lead, you must show up and do the work! I have heard this trait referred to by many names, but the most interesting is the "ministry of being there." That description has always stuck with me, and it makes a lot of sense when you think about the process. I have spent many hours in a hospital with members who didn't even know I was there advocating for them. Being there (showing up) is a very important aspect of leadership. It is difficult to misinterpret as anything other than being involved and connected to the people and the work. Being present at these critically important times makes a world of difference to those in need, in both the organization and the community. I have a strong belief that you lead best when you lead close-up.

Another great benefit of simply being there is that it will change some people's behaviors and attitudes. Your presence will lead them to be more productive. For example, I have attended dozens of community fire and life safety educational events. The members and I usually discuss how and why they should perform their duties to leave the best impression possible with our customers. I have found that this discussion is a very powerful process of street-level improvement that creates acceptance in members who may see themselves as "stuck" doing public education. I can recall one specific public education venture. We didn't have any citizens express an interest in fire prevention, so I challenged the crew working this event to stir up some interest by asking folks to stop by the Child's Fire and Life Safety House. Before long, the team had a line of children waiting to "stop, drop, and roll" and slide down our "ladder tower" blow-up kiddie slide (fig. 1–1). What a great improvement! Just because a leader showed interest, gave guidance, and directly supported the members by being there.

Fig. 1–1. "Ladder Tower" kiddie slide

MANAGEMENT BY WALKING AROUND

There is a popular national television show where the boss works with the line employees for a period of time. One episode was very interesting to me. During this episode, several unmistakable leadership issues occurred that are worth pointing out in this text. The one that immediately comes to mind is the segment where a cruise ship chief executive officer sets sail for a week-long onboard excursion with the frontline workers aboard the *SS Money Maker*. The boss wore the obligatory cheesy disguise to hide his real identity from the company's frontline workers. All goes swimmingly (pardon the pun) until the undercover CEO is assigned deck duty.

This newly assigned line worker helps the regular crew set up a simulated ice-skating rink for the guests to enjoy. To the boss's surprise, none of the guests use the skating facility, even though it takes a lot of crew time to set up. After a question or two to the crew, the CEO learns that no one really uses this expensive customer attraction. Without this personal, first-hand experience, the CEO would most likely be back at headquarters, potentially ordering the next seaworthy floating city with two simulated ice-skating surfaces.

The purpose of sharing this story is to clearly point out that the best way a leader can (and in most cases the way that they should) learn information about the organization and the current performance is to simply take a look for themselves. Remove as many of the filters between the leader and the people and work as possible. The best leaders have a very healthy curiosity. Bosses (at every level) need to have an inquisitive nature and should always be looking at what is going on in their outfit (the people and the work). Be aware of everything that is going on in the organization, within reason (some departments are very large). Again, the label micromanager may get thrown around, but never forget that you are paid (career fire-rescue officers) or elected (volunteer fire-rescue officers) to do just that: to look, be involved, and take a sincere guiding interest. Think of the dozens (maybe hundreds) of improper and embarrassing issues that are reported about public safety agencies in a one-year window. I would be willing to bet a small fortune that the leaders of those departments would love to have a do-over (or a mulligan in golf terms) where they could go back and be present to redirect or guide the involved parties to prevent the negative behavior that caused the organization stress and embarrassment.

Another benefit of managing by walking around is that your members will get into the habit of following your department's standard operating procedures (SOPs). Over the years, I have worked in departments that followed written guidelines a little closer when a chief officer started the response to an alarm.

I can remember as a battalion chief's driver being directed to continue a response to a "food on the stove" report. The chief had me drive to the rear of the structure to ensure that the second-due company officer had laid a supply line there, as required by policy. The previous chief never seemed to check up on a company's position at such small incidents, so sometimes SOPs weren't followed as strictly. As word got out that the newly appointed battalion chief was checking the companies assigned to the rear positions, even on small incidents, the required policy was reinforced and followed more closely.

BE A CHEERLEADER

When the leadership team shows up and everything is going well, then become the department's cheerleader. Let the hard-working and effective members know just how much you appreciate their great work. A simple "thank you" will go a long way.

BE A HOT COACH

At times, the leader may need to add the process I call "hot coaching": giving the member leading the work unit discrete advice or hints to make them highly successful. Sometimes our great people will simply forget to take care of something. Perhaps that *something* is a very important job task. For instance, I was operating at a significant house on fire. I arrived just after the first-in engine company. I was able to assume command before the second company arrived on location. I set our structural operations policy and called for a supply line, attack lines, exposure lines, ventilation, and utilities controlled. I obtained a 360-degree view and called for the rapid intervention crew (RIC) and the compressed air unit to respond as well. From my viewpoint, I was rocking the command position.

However, I failed to establish a search-and-rescue (SAR) group to check this 2,000-square-feet split level. When the shift commander (who was the deputy fire chief) arrived, he reviewed my operations by asking me a few questions. I was too busy to deal with his simple questions; I had a fire to put out. But the senior chief quickly realized that I had missed this mission-critical function.

Instead of blowing his stack or otherwise showing his disappointment, he just asked me to get a primary all-clear. I grabbed the radio microphone and gave the search assignment to the newly arrived second-due ladder truck. Because of this great boss, I looked like I was on top of everything and we never spoke about my shortcoming again. I bet I was the best SAR battalion chief in our system for quite a while. Hot coaching is a powerful and important skill for an effective leader to have.

LAST RESORT: ASSUME COMMAND

Let me be very clear: if you need to assume command of an event immediately by relieving a subordinate member, by all means do that exactly. I would submit, based on my many years of experience, that this is a very seldom-used process, but sometimes it is necessary. In such circumstances, swiftly and confidently take the action necessary to correct the problem at hand or to strengthen the leadership process if needed.

Sometimes things just don't go right (at an emergency or doing routine work). The window to be a cheerleader has passed and hot coaching is no longer a viable option. It is time to formally transfer command and take responsibility. There are times when a new supervisor will ask directly for a higher-ranking person to assume command of a specific situation. The higher-ranking boss must be ready, willing, and able to manage the situation at hand.

By always being there when you should, you will be casting a long shadow; everyone will work to not let you down even when you can't be in attendance (vacation, meetings, health issues, etc.). So, please: show up! You need to be present to be followed and to be a leader of your members. The best leaders on earth want to have both the minds and hearts of all of their followers. Seems to me, you have to be up-close, personal, and present to reach this level of followership.

SUMMARY AND REVIEW

To make this a more valuable and powerful document, this information can be applied to all aspects of your life. This book presents great information for the workplace, and the rules can also be applied during volunteer work that one may engage in. The sound, personal, effective leadership principles displayed on the pages of this book can be applied to just about any setting that requires you to take on the role of a leader.

When you are in a leadership position (or attempting to become a leader), your absence is deafening. The expectation is that the leader be an engaged part of the team. The best leaders are the ones who are close to the people and the work. They are often seen by their followers and know what is going on in their organization. The belief is that the "bosses" (lieutenants, captains or chiefs) who don't show up feel like the job is below them. This is a bad situation and can easily be avoided by simply showing up. Realistically, the leader cannot be everywhere all of the time, but when you can, be there with your frontline workers! These workers will quickly get a sense of whether the boss is willing to be involved and part of the team or aloof and unwilling to leave the comfort of the office.

CRITICAL LEARNING POINTS

1. Good leaders are always curious about the people and the work at hand. Be nosy and determine what is happening under your command.
2. Always be present if you can. Sending a representative is not the same as you actually being there. You will never know the reality of the situation unless you show up and learn first-hand, and you cannot correct actions or behaviors that you are unaware of.
3. It has been my experience that the simple presence of a leader can lead to a change in the behaviors and performance of the followers.
4. Be a workplace cheerleader. Thank the folks who are doing a good job for you! A simple thank you at the event will be appreciated, if your timing is appropriate and your words are sincere.
5. Use a hot coaching method. The idea here is to be a very clear and guiding presence to someone who may need a little help to get the work exactly right. If you take the time and make the effort to guide individuals who are not performing at the highest level, you will be remembered as a good leader and person.
6. When the operation is failing and the work is going south, do not hesitate to take over and resolve the related issues. But don't forget that the way in which you relieve another member will be remembered and talked about for a long, long time. Be professional and appropriate.

LEADERSHIP DISCUSSION QUESTIONS

1. List and discuss at least two situations where a leader failed to show up and things didn't go very well.

2. Two case studies are listed in this chapter describing instances when fire-rescue departments did not properly do their job. What do you think the community's reaction was to these failures? What would you guess as the root cause of these failures? What are five steps to overcome these types of problems and regain the trust of the public that we serve?

3. Describe at least two situations when the leader was present and the work at hand went well. Was the leader seen as a micromanager? Write a paragraph that speculates on what might have happened if the leader had not made an appearance at this event or emergency.

4. Give at least two examples where the leader became an impromptu cheerleader in the workplace. Was that behavior effective? Why or why not?

5. Give at least two examples where the leader became an impromptu hot coach in the workplace. Was that behavior effective? Why or why not?

___

___

___

6. Describe the circumstance that should exist when the leader relieves or otherwise removes someone for an important job. What are the benefits of taking this action? What are the risks of relieving someone from a command position?

___

___

___

PERSONAL LEADERSHIP PLAN FOR IMPROVEMENT

1. What behaviors do you demonstrate that are blocking you from showing up at various events with your department?
2. What behaviors must you adopt in order to always show up at various events with your department?
3. List at least two training programs you will attend in the next year to become a better leader for your department.
4. List at least two leadership books that focus on showing up that you will read in the next year to become a better leader for your department.
5. List the names of two members you will ask to be your mentors on the leadership skill of showing up.
6. List the names of two members of your department you will mentor on the leadership skill of showing up.
7. Add one other leadership commitment you are willing to make in the next 12 months.

PRACTICAL APPLICATION AND RELATED CASE STUDIES

CASE STUDY 1

The following story was reported by KCBS's Holly Quan. Reprinted with permission of CBS San Francisco.

> The Alameda Police Department Wednesday released 911 calls and a timeline of events surrounding the death of a man who intentionally drowned at Crown Beach on Memorial Day while rescuers watched from the shore for almost an hour, prompting outrage from the public.
>
> "He's trying to drown himself," 53-year-old Raymond Zack's elderly mother told a 911 dispatcher in the calls released today. "Hurry up, he's way out there. He doesn't swim. Please hurry."
>
> Zack had waded about 150 yards into the water near the 2100 block of Shoreline Drive. He had tried to commit suicide before, his mother, Dolores Berry, told the dispatcher at 11:33 a.m.
>
> Emergency responders arrived on scene just minutes later, but they watched from the beach as Zack eventually lost consciousness and was brought to shore by a good Samaritan, according to the police transcript. Zack later died at a local hospital.
>
> Fire department protocol prohibited rescuers from going into the water because their water-rescue certification had lapsed, the fire department's acting deputy chief of operations, Daren Olson, said the day after the drowning.
>
> According to the transcript, police contacted the U.S. Coast Guard at 11:32 a.m., two minutes after the first call was made to 911, to request a rescue boat and helicopter for Zack.
>
> The Coast Guard said its crews were about 40 minutes out, so the Alameda Police Department contacted the sheriff's office, Oakland Police Department, and Alameda County Fire Department in search of a closer boat.
>
> None of the agencies could offer faster assistance, according to the transcript.

At 11:42 a.m., the Coast Guard said its boat was about 15 minutes away and almost ready to leave. At 12:06 p.m., the agency advised police its boat and helicopter were en route, but by then Zack appeared to be floating face down, according to the transcript.

At 12:17 p.m., the Coast Guard radioed police to say they could not reach Zack because the water was too shallow. Alameda police then called the East Bay Regional Park District to see if they had any crews available.

Around the same time, at 12:24 p.m., police instructed dispatch to ask Alameda County Search and Rescue to send someone out to do a "recovery." At 12:27 p.m., the officers on the scene radioed dispatch to say a volunteer would go into the water to retrieve Zack.

Police then called the dispatch center to request crime scene screens for privacy once Zack was brought to shore.

It turned out the Coast Guard helicopter was occupied on another call and needed to refuel, and the Police Department does not appear to have called the California Highway Patrol to see if a chopper was available.

The day after the drowning, the fire department changed its policy so rescue swimmers could be sent into the water at the discretion of the incident commander on scene, acting city manager Lisa Goldman said at Tuesday's City Council meeting.

The department has 30 volunteers scheduled for rescue swimmer training, including 16 who will begin the certification process next week, she said.

The city will also conduct an independent review and is compiling documents relevant to the incident, Mayor Marie Gilmore said at the meeting. The materials will include transcripts, timelines and department memos regarding water-rescue training.

The site was under construction today and is expected to go live by the end of the week, the city manager's office said.

"This has been a really tragic, tragic situation for obviously Mr. Zack, and certainly his family and members of the public," Gilmore said at Saturday's meeting.

"We will be as transparent as we can," she added. "We understand that there is a competence problem right now, and we want to assure

> our residents we have made and will continue to make the changes that are necessary."
>
> Deputy Fire Chief Olson said last week that if a firefighter had disobeyed protocol by going in after Zack without proper training, the individual would not necessarily have been punished.
>
> But Zack was an adult man, 6 feet 3 inches tall and 280 pounds, who was intent on taking his own life, Olson said. First responders had no way of knowing if he were armed or would have tried to hurt a would-be rescuer.
>
> "Grabbing an arm and tugging him to shore is not only maybe not an option, but perhaps recklessly dangerous," he said the day after the drowning.
>
> The water was also about 54 degrees, adding to the difficulties rescuers faced, Olson said. (Quan 2011)

CASE STUDY 2

In this chapter, I referred to the case where a fire company had a customer in need of assistance right outside their door and did not show up. The result was a loss of life. An internal investigation was completed on the unfortunate incident, and it is clear that there was a lack of leadership and accountability that day. The leader did not show up; he simply did not care. The following is excerpted from a *Washington Post* article.

> A series of miscues, poor communication, and outright apathy led five firefighters to ignore desperate pleas to treat an elderly man who had collapsed across the street from a Northeast Washington fire station and later died of a heart attack, according to an internal report, made public Friday, urging that all five be disciplined.
>
> Three firefighters who were in the station's kitchen and a veteran lieutenant heard two announcements by a just-hired firefighter but didn't immediately react. When one firefighter finally sought out the station's lieutenant, instructions were unclear or dismissed. Instead of rushing across the street to help, the firefighter retired to his bunk bed to study for a promotional exam, the report produced by the mayor's office says. (Hermann 2014)

This incident is a prime example of what can happen when leadership fails. While there are several news articles, blogs, and social media comments regarding what happened the night of January 25, 2014, the common thread is, "How could anyone, particularly a first responder, just ignore someone in need?" By not showing up—taking responsibility for the situation and caring about the work—the lieutenant allowed the death of a customer, someone relying on the department for help. This is unconscionable and should absolutely never happen.

2

Lead from the Front

In reading the lives of great men, I found that the first victory they won was over themselves... self-discipline with all of them came first.

—Harry S. Truman, 33rd President of the United States

The second leadership rule relates to being a courageous leader. A courageous leader is one all members willingly follow. The phrase "winning hearts and minds" is brought to life by a courageous leader. Great leaders develop followers who are willing to be led without question or without exception.

Leading from the front is a simple, straightforward behavior typically accomplished with self-discipline, training, education, competence, common sense, and wisdom. This rule is a concept that is easy to understand and easy to write about; however, at times, it is difficult to implement and live up to. Interestingly, it seems that it is more difficult to lead by example for long periods of time—such as the length of a fire-rescue service career of 25 or 30 years—than for a few hours or even a shift cycle. With this in mind, leading from the front becomes increasingly important to discuss, understand, and research in order to help you be a successful leader for the long haul.

The military focuses on this same principle of leading from the front. The motto of the US Army Infantry Branch is simply "Follow Me." The army's infantry officers are trained and prepared to lead their men and women into battle. If a person is asking their members to follow them, that person must be self-motivated and always willing to lead from the front, to give guidance and direction to all of the members of their unit, company, battalion, or division according to their rank and position within the agency.

Many great leaders are developed understanding and using the "follow me" process. This rule may be an informational review for some and perhaps a brand-new process for other readers. Either way (long-time leader or brand-new boss or somewhere in between), this is a mission-critical step, as all of these leadership rules are necessary in becoming and staying a top notch and well-respected fire department leader.

When the boss leads from the front with a positive attitude and a great operational skill set, these three traits become a very powerful mixture. The most effective leaders come packaged from the factory understanding these aspects (leading from the front with a positive attitude and being able to do the work at hand).

Never lose sight of the fact that every leadership position comes with a tremendous level of organizational scrutiny and visibility. The people who want to step up and lead should be ready for (and actually welcome) the highest level of scrutiny and visibility while conducting their job assignment as the leader of the team. Scrutiny and visibility rightfully come with the mantle of being the one who pins on the gold collar insignias with those funny-looking speaking trumpets. (In the nineteenth century, the fire service started the practice of using speaking trumpets to disseminate orders and manage the fire ground. When the walkie-talkie radio made an appearance, the speaking trumpet was relegated to a place of honor to indicate a person's fire-rescue service rank. So, those symbols worn on an officer's collar are not bugles. Always use the correct terminology for everything in our profession, whenever possible.) Since leaders are watched closely all the time, the need for competence and positivity about the department are obvious.

There are epic examples of folks saying one thing while doing exactly the opposite. Do you remember the arson investigator John Leonard Orr? He was a highly decorated fire service professional who seemed to be on top of his fire investigations game in every aspect. He was a very effective and well-respected fire captain in Southern California who was revered by many, and was a sought-after lecturer and author on the topic of fire and arson investigation. Little did the world know the real character of this man.

The country and the fire service was amazed to find out that Orr was one of America's most notorious and destructive serial arsonists. He was described by the Bureau of Alcohol, Tobacco, and Firearms (ATF) as the twentieth century's most prolific serial arsonist. Orr was found guilty of setting dozens of fires up and down the West Coast and was found responsible for several associated fire fatalities. As Orr would travel to deliver lectures on the topic of investigating arson fires, he would stop along the way and set significant buildings on fire. Then, Orr would *just happen* to be in town after these major events, when he would offer his services to the local fire marshal. He seemed to do the nearly impossible by determining the cause of the fires (the ones he'd set just a few hours earlier) quickly, accurately, and effectively. After being found guilty by a jury of his peers, Orr is now in a California state prison serving a life sentence plus 20 years without the possibility of parole for multiple counts of murder and arson.

Most fire department leaders, from firefighters to fire chiefs, prepare and circulate a wealth of written documents (memos, emails, member evaluations, reports, and the like). Every member of the organization (and in certain circumstances where public information laws apply, the public) can read or hear what you are saying. Your written words have been captured (in some form or another) for all of time and your actions (performance) must agree with what you say. Always walk the walk if you talk the talk. Be true to your core values, ethics, and your oath of office. Align your words and actions every single day. You will be respected and trusted for living out that leadership principle.

Leaders understand that someone is always watching them! Everyone in your agency is watching you as you lead. The members in your outfit (under your supervision or command) hear the words you say, but your actions are what they see, believe, react to, and remember. To be a courageous and effective leader, your words and your actions must line up and be synchronized. As a leader in your department, there can be no difference between what you say and what you actually do. The best advice that I can give to you is to determine what your core values are and stick with them through thick and thin. "Don't tell me what you can do, show me what you can do" is the prevailing fire service opinion.

One of the most impactful leadership books that I can recommend is General Colin Powell's *My American Journey* (2003). The past chairman of the Joint Chiefs of Staff writes about the fact that someone is always watching. Powell was always under great scrutiny and review: his work was carefully examined by many others who were in higher- (and for that matter, lower-) ranking positions. The chance to be a successful and a trusted US Army leader was his motivation and focus every day. General Colin Powell passed the leadership test with flying colors and went on to an amazing military career. With his simple words, he reminds us all that someone is always watching us perform our duties. I would add that there is more than a passing interest in public safety officers' off-duty behaviors as well.

LEADERSHIP STYLES

Many years ago, I came across an undercover news program created by a local TV network and titled *Men at Work*. The program compared public works departments in three communities performing such duties as street repairs and litter removal. Two performed above the viewers' (and media's) expectations. The news reporter raved about the workers, the supervision, the management performance, and the government employees in general. Undercover reporting in the third community, however, left viewers wondering whether any work was performed by the public works department or, for that matter, by any of that city's employees. It was an indictment of all government employees: the program showed the employees not caring, not performing their required duties diligently, and violating the public's trust by loafing instead of working. One supervisor was caught on tape littering and even relieving his bladder in public on the very street he was charged with keeping clean. What were they thinking?

There is a very interesting side story embedded in the undercover report. Each of the three agency heads demonstrated a very different leadership style. The boss in the first community used the democratic leadership style, where he shared the decision-making with his rank-and-file employees and created buy-in for them. The director was often seen among the workers asking for specific tasks to be completed while providing support and obtaining constant employee input.

The second leader profiled was an autocratic manager who demanded a very high level of work ethic of his personnel. The news story on his community opens with him reciting his work mantra: "Don't tell me what you can't do. Tell me what you can do." Early in this story, the same character is described as "the Bobby Knight of the public works world." In the closing scene of this story, the director states that if he was "given enough people, equipment, and materials, [he could] get a highway put into Florida by Monday." His focus was on getting the work completed by the employees. This public works executive used the very deliberate, directed decision-making process known as autocratic leadership.

The final leader was never reached in person. His work crew did not perform very well and obviously lacked supervision and accountability. The laissez-faire or hands-off style is negatively exaggerated in the way this gentleman managed his personnel. The closing four or five stories show the reporter searching for comments from this manager—the public works commissioner himself never appears on camera.

This very valuable undercover series examines the types of leadership that can be selected. Many people gravitate toward democratic leadership: get the frontline workers involved in making the decisions and encourage them to do their best at the job at hand. Other people tell me that the very best leadership style is the autocratic style: the workers are not confused about leadership's direction and are well aware of the goals and objectives for the workforce, which is sometimes the most efficient way to get the work at hand completed. Still others would point that the laissez-faire style is best because most folks nowadays do not like to be told what to do at work: they know what to do and how to do it, so why should the leader always be a micromanager?

To be sure, all three styles are important and have a place in properly leading folks. The trick is determining what style is most useful in what setting, because the various leadership styles are situational and can be effectively used based on the problem or issue. When the setting allows employee input, use the democratic style: perhaps the department is selecting a new fire engine for purchase. Allow members to have detailed input as to the brand, the features, the model, and so on. At emergency scenes or other operations that do not allow for discretionary time, the autocratic style may fit best: the department should not be taking a vote to lay a supply hose line and advance an attack line into a burning building. Finally, when the employees or members are familiar with the task at hand and have demonstrated success at handling the specific situations, the best choice may be the laissez-faire style of leadership.

The real magic of your leadership is knowing when to apply each of the styles. Each of us has a preferred leadership style. Mine is the autocratic process. When I interact with folks, I have to remember that and not slip into the trap of always relying on this particular style. Do not let your strength become a weakness; know when and how to use all of the basic types of leadership styles.

What would the results be if the local news media spent a day secretly following you and your crew for 12 hours? We must always think about how we appear and are perceived by our stakeholders (citizens and visitors). Always—and I mean always—lead from the front if you want to be a courageous leader inside your agency.

I have to mention the need to be a loyal member. All members need to be loyal to their department, to their bosses, and to each other to be successful leaders. When a person badmouths their boss or peer, or doesn't follow orders completely (or follows them begrudgingly), it undermines the leaders and peers, but also undermines the specific individual's ability to be a leader. When the leader speaks negatively about their superior or their peers, an uncomfortable situation develops that could have been easily avoided by speaking directly to that person in private or, if that fails, going to their superior, rather than openly sharing the acrimony with subordinates.

Now, this is not saying that you should follow a poor leader. Give the boss the benefit of the doubt and be a loyal follower until that respect and trust are lost. Once that happens, for the sake of your peers, your subordinates, your department, and yourself, you need to have the courage to speak with them about the issues you've noticed. This does not need to be a public cal-lout—in fact, it should not be—but mistakes and incompetence in the fire service can lead to job dissatisfaction, loss of funding, public outcry, and, in the worst-case scenarios, deaths, both of civilians and firefighters. If your superior is not willing to listen or is unresponsive, the conversation needs to be taken further up the line of command.

SUMMARY AND REVIEW

Always remember that someone is watching! Whether it is the public, the media, or your own people, you are being very closely watched by people who are paying attention to everything you do and how you do it. Make sure that your words and actions are in line with one another.

I was on a flight one time when an article in a national newspaper caught my eye. The headline was something about how the average American citizen's image is captured 22 times per day on a security camera. My guess is that if you live in a highly urbanized community (say New York City or Chicago), this number is much higher, and the reverse is likely true in small towns. Couple the security camera activations with the cameras embedded in pretty much every cell phone and it is clear that there are a lot of eyes on us. Behave properly at the fire station, at alarms, and even when you are off-duty.

Fig. 2–1. DC's last horse fire response, July 1925

CRITICAL LEARNING POINTS

Never lose sight of the fact that fire departments and all public safety agencies are highly visible and highly scrutinized. It is impossible to get away with inappropriate or illegal behaviors. Guide the department away from controversy and negative issues before they have an unfavorable impact. The best disinfectant is sunshine—keep all activities in the light. Be transparent in all that you do.

1. Never underestimate the value and power of a positive attitude. Formal, informal, and aspiring leaders always need to keep a positive

attitude about the job, themselves, and their subordinates. When the focus is solving a problem or doing the work at hand rather than complaining, the task is completed successfully and quickly.

2. Should bosses say one thing and do another because of the privilege of their positions or the belief that they will not get caught, their effectiveness as a leader is undermined.
3. Pick the correct leadership style (democratic, autocratic, and laissez-faire) for the situation at hand. All leadership types and traits are useful, learn how and when to use them. Maybe, more importantly, know when not to use a specific leadership style. Each of us has a preferred leadership style. Determine what that is (know yourself) and keep that information in mind all of the time. Do not let your leadership strength become your leadership weakness.

LEADERSHIP DISCUSSION QUESTIONS

1. Prepare a written description of 500–1,000 words about what "lead from the front" means to you.
2. Based on your response to question 1, have a discussion about "leading from the front" with your unit, company, battalion, or classmates.
3. Reflect on the fallen leaders in our society who did not align their words and actions. Most of them are folks who were once in powerful and prestigious positions. Don't be one of those lost leadership souls.
4. List each of the three primary leadership styles. Give a detailed description of each type. Provide two examples of times when each leadership style would be useful. Provide two examples of times when each would not be appropriate.
5. General Colin Powell wrote that "someone is always watching" in his 1995 autobiography *My American Journey*. What do you think that General Powell meant? What are some examples? How can this information be used in a positive and constructive way?
6. Provide your opinion on the statement "Public safety service is a highly scrutinized and highly visible service." If this statement is true, what is the impact of this? Positive? Potential negatives? What would happen if the news media spent the day with your unit undercover? What positive items would they observe? What negative behaviors would present themselves?

PERSONAL LEADERSHIP PLAN

1. What behaviors do you demonstrate that are blocking you from leading from the front with your department?

2. What behaviors must you adopt in order to always lead from the front within your department?

3. List at least two training programs you will attend in the next year to become better at leading from the front.

4. List at least two leadership books you will read in the next year to become better at leading from the front.

5. List the names of two members you will ask to be your mentors to learn the leadership skill of leading from the front.

6. List the names of two members of your department you will mentor to learn the leadership skill of leading from the front.

7. Add one commitment you are willing to make in the next 12 months in order to learn to lead from the front.

PRACTICAL APPLICATION AND CASE STUDIES

CASE STUDY 1

CASE STUDY: FIRE, EMS RESPONSE TO ACTIVE SHOOTER

Dennis L. Rubin

I worked my hardest to become and maintain National Registry Emergency Medical Technician—Basic (NREMT-B) certification right after I was appointed fire chief. I wanted to demonstrate the same skill set demanded of the members; leading by example is a long-time core personal value.

In the mid-1990s, I found myself impatiently watching the mail for my NREMT-B test results. Finally, the white tyvek envelope showed up. The hard work and effort paid off, the gold threaded shoulder patches would soon be added to all of my uniform shirts.

It seems like only yesterday that I was sitting in the front of the Fire-Rescue Training Center classroom at the Westgate Fire Station. The course instructors, Paramedic/Battalion Chief Larry Williams and his training staff, promised all of the students that if we worked hard and studied the textbook they would help everyone to get through the all-important final national registry examination.

After four weeks of class work, he was true to his word. The entire EMT class would be successful at becoming NREMT-Bs. I likely didn't thank the chief and his team enough for their efforts. So, I give public thanks to Chief Williams and staff—better late than never.

"SCENE IS SAFE"

Chief Williams and his staff insisted that we be able to perform every required NREMT-B skill flawlessly. Each time a student would be put through a skills simulation, that EMT candidate would verbalize, "body substance isolation techniques in place by all responders."

Next, the class members would take a 360-degree look around the staged emergency medical scene. Perhaps the chief added a downed electrical power line or a crazed bystander wielding a knife for the trainee to contend with and successfully resolve.

Once the area was cleared of obvious dangers, the student gave the "scene is safe." It was at this point, our practical skills instructor would respond that he/she copied that body substance isolation equipment was in place and the emergency incident scene was rendered safe to enter.

This two-step process was drilled into everyone who participated in this emergency medical training program.

THE REAL DEAL

Move the calendar up about 10 years. I knowingly participated in a major operation that broke the emergency "scene is safe" rule.

In fact, the incident called into question the long-standing wisdom about always making sure that the scene is safe before delivering patient care. As this event unfolded, it didn't seem like I would be engaging our personnel in this high-risk fashion.

It was Friday, March 11, 2005, just before 9 a.m. I was serving as Atlanta's fire chief. I had just completed attending a meeting at city hall when Paramedic Engine 1 provided an on-scene report of a man down on location at Martin Luther King Avenue just before Pryor Street.

Soon after, the fire company officer reported that there was a Fulton County Deputy Sheriff who had been shot, and advanced life support (ALS) protocols were being implemented. Being in close proximity to this alarm, I responded to see how I could assist.

We quickly learned that the deputy was fatally shot, but didn't know this officer was part of a much larger active-shooter incident.

ON-SCENE COMMAND

The on-scene sheriff deputies were asking for help inside the courthouse for persons who had been shot in one of the courtrooms. I immediately declared a major medical event and asked for a mass casualty response with four ALS ambulances.

Next, it was time to set up command on the trunk lid of my chief's car. The initial incident action plan was simple: ensure there were enough emergency medical resources to handle the victims who would soon be removed from the courthouse. We notified local hospitals to request the emergency departments prepare to manage multiple gunshot victims.

To complete the initial incident action plan (IAP) (figs. 2–3 through 2–5), there were other issues that would have to be resolved. In particular, this large incident site would require tracking on-location companies and their assignments to account for everyone at the event.

Five Elements of Hazard Zone Accountability

1. *Who are the firefighters/EMTs working in the hazard zone?*
2. *Where are the firefighters located in the hazard zone?*
3. *What actions (tactics and tasks) are they performing?*
4. *What are the current and forecasted conditions in the hazard zone?*
5. *What are the two paths of exit from the hazard zone?*

Fig. 2–3. The incident command and accountability officer must be able to answer these five basic accountability questions when firefighters are placed in a hazard zone.

Initial Incident Action Plan

Courthouse Shooting

1. *Provide a brief initial report.*
2. *Establish command.*
3. *Ensure that Engine 1 has the resources to assist the deputy sheriff that was shot out on the street.*
4. *Declare a major medical emergency—request emergency medical help.*
5. *Notify receiving hospitals of the situation and potential of the upcoming patient flow*
6. *Determine an appropriate staging area—keeping roads open.*
7. *Account for and track operating companies.*
8. *Identify a PIO to work with the news media.*
9. *Notify City Hall executives.*

Revised & Final Incident Action Plan

Courthouse Shooting

1. *Provide situation status report.*
2. *Liaison and communicate with all of the law enforcement operations.*
3. *Designate a hazard zone entry medical response team.*
4. *Provide treatment of the sick and injured victims.*
5. *Ensure rapid transport to hospital facilities as needed.*
6. *Provide for decontamination of personnel.*
7. *Document the response to the event.*
8. *Participate in a live press conference with Mayor Shirley Franklin and police agency officials*
9. *Conduct a hot wash.*
10. *Critical incident stress debriefing as needed.*

Figs. 2–4 and 2–5. Incident action plans for the Fulton County Courthouse shooting

This incident would also require command to establish a public information officer to keep the mayor's office informed of the fire rescue department's operations and to help with the media as requested. And it would be necessary to stage incoming units in an area that was safe for the evolving law enforcement operation and not blocking ambulance access or egress.

ALL GREAT PLANS

Firefighter/paramedics and the necessary ALS equipment were positioned in front of the courthouse and in close proximity to the command post. My plan was that the serious to critically injured would be removed from the building and the prehospital care teams would be assigned a patient, an ambulance and on their way to the hospital as quick as possible.

The planned was simply not happening. No patients were being removed from inside this very large judicial facility.

About then, Dr. James Augustine, our medical director and assistant fire chief, arrived. I was quite pleased to have him at the command post [and] asked him to validate the IAP. Dr. Augustine concurred and we intended to move forward with the plan.

A Fulton County deputy sheriff was assigned as liaison with the fire command post. The details that the deputy provided were not very promising.

SPOTTY INFORMATION

The belief was that the sergeant who was murdered on the street next to the courthouse was in pursuit of a criminal who had escaped while being transported from the jail to the courtroom for this day's proceedings in a lengthy criminal case.

When asked if the alleged perpetrator was gone from the scene, the answer was a chilling, "We are unsure." We asked more size-up seeking questions [of] this officer.

- Was the shooter operating alone?
- Were long weapons or automatic weapons involved in the shootings?
- Was there anyone else shooting inside of the courthouse?
- Has the shooter(s) inside the courthouse been neutralized?
- What are the locations of the people shot inside the building?

His answers to all of our questions: "We are unsure." The officer did say that a SWAT unit was being assembled in front of the building.

The potential life loss caused by the delay in reaching and removing the injured was a real possibility. Dr. Augustine strongly suggested that he should travel with SWAT into the hot zone to triage, treat, and remove the shooting victims.

GOING IN

I reluctantly agreed to his heroic request. Dr. Augustine was assigned several firefighter/paramedics who volunteered to go into the hazard zone.

In minutes police made entry; immediately trailing SWAT was Dr. Augustine and two fire fighter/paramedics with a small amount of medical equipment. As each floor of the courthouse was systematically searched and swept for victims, the doctor gave a detailed radio situation report to command. An operational line drawing of the building was updated at the same time.

The news from inside the building was not very good. There were two fatalities—a judge and a court reporter. Another deputy was in critical condition from a reported gunshot wound—it turned out the officer was beaten severely, not shot.

The deputy was treated and rapidly transported to a receiving facility where she survived. Another patient suffered severe chest pains. He was provided with prehospital advanced cardiac care treatment and transportation.

Finally, there were several other minor injuries. Those simple bruises and cuts occurred in people who were running to avoid being shot.

One major step that was missing, and I take full responsibility, was providing post critical incident stress debriefing of the members on location with emphasis on those who entered the hazard zone.

QUESTIONING LONG-HELD WISDOM

I had serious second thoughts sending that team into a highly hostile and volatile dynamic situation. The only protection that my team was afforded was the coverage that SWAT members could provide.

As it turned out, quadruple murderer Brian Nichols had escaped the scene and was a lone gunman. Of course, this was not known at the time of the entry into the hot zone.

After years of training prehospital care providers to ensure that the scene is safe, that wisdom is now being questioned.

The US Fire Administration reports that the average active shooter event ended within a few minutes. The Fulton County courthouse shooting in 2005 followed the average.

Generally a shooting victim's injuries are critical to life-threatening. The active shooter will likely use a high-powered, large-caliber automatic or semi-automatic weapon, causing critical penetrating wounds and hemorrhage. Evidence-based emergency medicine mandates immediate prehospital interventions.

The bleeding must be controlled or stopped. The patient's airway has to be opened and maintained as well. Fluid replacement may be needed quickly to restore perfusion. Finally, rapid transportation to a receiving hospital—preferably a shock-trauma center—will be required. Once there, surgery is likely.

HOW TO SAVE LIVES

Armed with this information, it is clear that if there is any chance of saving human lives at an active-shooter event there are three mission-crucial steps that we have to provide.

- Hemorrhage must be stopped.
- Rapid removal to the treatment area for immediate prehospital care.
- Expeditious transportation to a trauma care facility.

Any other action plan steps will not work to have viable victims with a chance of survival from these frequently occurring events. Waiting until after a scene has been secured to treat victims could be an impediment to saving lives in situations like these, where every minute matters so greatly.

If on-scene emergency medical care is withheld until the scene is safe, victim outcomes are clearly compromised. This emerging situation raises many questions, challenges and of course, opportunities.

The questions that must be answered are fairly simple to formulate, but very difficult to answer. Here are 10 of those hard questions.

- Do we continue to use the mandate of scene safe or don't enter the area?
- Is there a place for prehospital care givers to enter an active-shooter hazard zone?

- Do we travel with SWAT during the entry and search for victims?
- If EMS goes in to the hazard zone, what additional training is needed?
- What personal protective equipment will be needed (bulletproof vests, ballistics helmets, weapons, etc.)?
- What certifications and authorities (limited police powers) are necessary to perform this service?
- What are the recertification cycles for training?
- What is the replacement cycle for the protective equipment?
- What type of pay increase is appropriate?
- Is participation in this type of program voluntary or mandatory?

Like military field medics A program called Counter Narcotics and Terrorism Operational Medical Support, tactical medic for short, prepares paramedics to travel with SWAT units to provide care for injured public safety members. The framework was similar to a military field medic.

Perhaps the tactical medic program could become the curriculum for the active shooter hazard zone entry process. Of course, the mission will be expanded to care for public safety officers and civilian victims.

Looking at the devastating active-shooter events in places like West Webster, N.Y. and Newtown, Conn., we need a different approach. If paramedics have been making tactical medic entries for years to protect police officers, we need to seriously consider expanding this program to add the growing numbers of victims of active-shooter violence.

Getting a verbal "scene is safe" before proceeding is still critical on most calls. But changing times and tactics make us rethink the absolute nature of that rule.

Dennis L. Rubin

MEMORANDUM

Series	Number	Originating Unit	Effective Date	Expiration Date
2009	**126**	**OFC**	**August 10, 2009**	**N/A**

Subject: **Employee Responsibility**

"The Past Is The Past"

It is important for all members of the Fire and & EMS Department to remember that the foundation of all we do is built on public trust. The taxpayers demand and expect public safety representatives to possess the highest moral character and integrity. The public does not see or understand a distinction between "off duty" behavior and "on-duty" behavior as it relates to anyone entrusted with their safety and security. Without the public trust, the Department loses the support of the community. The success of our mission is not wholly dependent upon operational efficiency but is built upon those critical pillars of public trust, integrity, truthfulness and ethical behavior.

Maintaining the public trust is paramount; past bad practices whether real or imagined, will not hold the Department hostage. This means that how previous administrations dealt with these types of issues is not the responsibility of the current administration. We will make decisions based on objective standards while at the same time recognizing that the specific facts of each incident must be examined on a case by case basis, in accordance with public law and regulation, and the General /Special Orders of this Department.

The Department understands the importance of developing and implementing a fair and equitable disciplinary process. The rules must be clear and unambiguous; the process must be known and understood by all, the penalties will be consistent within the circumstances, and the process must be timely. To ensure past organizational lapses don't occur in the future, the Department has been transforming the process by proactively making changes in the manner in which we prefer charges, propose and impose discipline, including establishing a single Chief Officer at any one point in time to handle chief's conferences. These changes have begun to produce greater consistency and transparency throughout the Department.

Programs such as random drug tests and annual background checks are a critical part of the overall process of maintaining public trust. It should be the hope of all employees that we work in a drug-free and violence-free workplace, where rules and expectations are understood and respected, and where self-discipline is the only discipline needed.

Dennis L. Rubin
Fire & EMS Chief

3

Flawlessly Execute the Basics of Your Job

Almost all quality improvement comes via simplification of design, manufacturing layout, processes, and procedures.

—*Tom Peters, American businessman*

In this chapter we discuss the leader's obligation to ensure that the job at hand (fire, emergency medical, hazardous materials mitigation, rescue, etc.) is completed effectively, efficiently and safely.

I love the notion of having great a customer service program embedded in every way in our departments. This concept was pioneered roughly 30 years ago by Chief Alan Brunacini from the Phoenix Fire Department (AZ), and will be covered in detail in chapter 9. However, we must resolve the emergency at hand quickly and professionally; no ifs, ands, or buts accepted. Resolving the emergency quickly is the number-one priority of any response system. Let's face it: everything gets much better after! Before the incident commander calls the communications center to get the customer service ball rolling, the incident must be under control. If the fire is still advancing and out of control, but the operational focus is contacting the local Red Cross or Salvation Army, the strategic plan is falling apart and out of balance. Obviously, Chief Lloyd Layman's "Big Seven" structural firefighting strategies (fig. 3–1) must be satisfied or at least well underway before related support activities can be implemented. Put the fire out and get the incident under control as quickly as possible!

Chief Lloyd Layman's Strategic Objectives

RESCUE	VENTILATION	SALVAGE
EXPOSURES		
CONFINEMENT		
EXTINGUISHMENT		
OVERHAUL		

Fig. 3–1. Chief Layman's strategic objectives (RECEO-VS)

In order to perform our sworn fire and emergency medical duties, we must be capable of doing the work at hand correctly the first time, every time. At a minimum, firefighters and firefighter emergency medical technicians (EMTs) must be certified to national standard skill levels.

The axiom is, "Extinguish the fire and all conditions will improve quickly." Of course, you cannot always apply this tactic immediately, but it is a good one to keep in mind. The tactics of extinguishment will be based on the situation. Factors such as time (preburn and operational set-up time), the fuel type and configuration, building construction, and so on are all items that will determine how long it will take to extinguish a building fire.

Nothing can ever replace having the correct skills, knowledge, and abilities to handle any emergency, regardless of whether it is a multisystem trauma or a second-alarm apartment fire.

As a big fan of the US military, and the US Marine Corps in particular (fig. 3–2), I have a wonderful phrase: "Every Marine a rifleman."

Fig. 3–2. "The Few. The Proud. The Marines."

The point is to focus attention on the core mission of the Marine Corps. To protect our nation from foreign and domestic enemies, the US Marine Corps must have excellent infantry skills. In the Marine Corps, whether you are a four-star general or a newly appointed private, first you are a rifleman. The comparison is that if you are going to claim to be a firefighter (regardless of your affiliation—volunteer or career—or your rank), you must maintain the basic training requirements and certifications to do your job (e.g., NREMT, Firefighter I, Firefighter II).

During a classroom presentation on the topic of leadership, a slide appeared bearing this Marine Corps phrase. A chief officer raised his hand to comment. He came to his feet (a little unusual for a laid-back fire-rescue leadership class) and went on to tell an interesting personal story. This fire chief was a "ring knocker" (military academy graduate) from the US Air Force Academy in Colorado Springs (CO). He had distinguished himself in his studies, graduated with honors, and received a commission as a second lieutenant in the US Air Force. The first official act of this new "butterbar" lieutenant (a reference to the gold-colored bar indicating his rank) was to request the transfer of his commission to the US Marine Corps. After a few weeks of waiting and a few other administrative steps, he was accepted into the Marine Corps. The chief thought he would be immediately commissioned as a second lieutenant in the Marine Corps, but not so fast: his first assignment was to be shipped off to Parris Island (SC) to learn to be a rifleman, which was followed by the standard officer candidate school. There were no shortcuts and the experience gave him a good measure of respect for the core value of being a Marine (rifleman).

The fire leadership class member said that he retired from the Marine Corps 25 years later as a lieutenant colonel before entering the fire service as his second career. In each of those 25 years as one of our American warriors, he emphasized, qualifying with a rifle was a standing requirement. This story reinforced my mission-critical leadership rule and continues to keep me focused on just how important it is to flawlessly execute the basics of our jobs. Let's face it, someone's life (including your own) depends on you embodying this rule.

The measurement for all of our credentials (fire, rescue, hazardous materials, and the like) should be the same as our prehospital care emergency medical certifications. I have had the wonderful opportunity to be a part of several fire-EMS departments in different states over the past few decades. With each opportunity (new department), the state health department (or the equivalent) would review my EMS records and prescribe the training and updated certifications I would need to be an EMS-participating member of the new outfit. Whether it was a required CPR recertification course or a completed NREMT-B course and national testing, it was always clear what I had to do to be qualified and certified to deliver prehospital care in that community. For my fire-related certifications, I was the final authority in determining whether these were acceptable. There was no outside department reviewing my certifications.

As you can probably tell, I am very much in favor of national certifications for all of our core disciplines. We must use the various national certification agencies (NFPA, IFSAC, etc.) to ensure that we have reached the minimum training and certification standards needed to do our jobs. Our fire duty skill set (initial and on-going) certifications should mirror the national registry emergency medical technician training and validation process.

Once we have met the minimum national training and certification standards, we must continue improving by reaching higher certification levels and capabilities. That is raising the bar as the leader (be it the formal or informal leader of the team) to make sure that you and your department members grow and keep up with the changing times.

I can remember back in 1971, Private D. L. Rubin was required to successfully complete the American Red Cross basic and advanced first aid courses as part of a nine-week firefighter recruit school. This training was great for the times, but within a year, it was determined to be outdated and not comprehensive enough for emergency ambulance service. Our department adopted and required emergency medical technician certification as the baseline training to deliver this service to the citizens and visitors to the county. Fire-based EMS agencies have never looked back. Today's standard baseline includes NREMT-B as a minimum requirement.

Can you imagine not obtaining current training information? This could never happen in the prehospital emergency medical care world. I am asking you to make the same commitment in all aspects of your career, not just emergency medicine. The National Fire Academy offers one of the best leadership development processes available with the Executive Fire Officer Program. This four-year training curriculum touches on all types of strategies to improve your effectiveness as a leader. You should also consider obtaining the Institute of Public Safety Excellence's Chief Fire Officer Designation and Chief Medical Officer Designation, to name just a few professional attainments. These designations are a specific and measurable way to determine whether you are operating at the top of your game. You must keep your skills, knowledge, and abilities up to national standards, no question.

There is no more important duty requirement than for a firefighter or firefighter-EMT to be able to flawlessly perform the basics of their job. When the chips are down and the customers are in need of help, you will always have your training to fall back on to resolve a difficult problem. The lives of all first responders and the lives of those in the community they are sworn to serve are literally on the line, just about every time an emergency vehicle goes out the door.

SUMMARY AND REVIEW

Flawlessly executing the basics of your job should be a standard tool in your leadership toolbox. Bosses of all types must support their subordinates and ensure that everyone under their command can reach this benchmark level and perform the needed services at emergency and nonemergency events. Skill-refresher training for technical hands-on skills needs to be reviewed on a regular basis. It is easy to lose a skill set that is not often used, and it is difficult to remember the exact steps of skills without refresher training. Most experts will point out that every day should be a training day in a progressive fire-rescue department. I fully agree. In order to deliver the "good goods" when our services are needed, we have to train and train often. I would suggest that four hours per shift (career members) or weekly duty crew shifts (volunteer members) need to be dedicated to a comprehensive training process as the bare minimum.

The notion that members need to be lifelong learners must be covered in a discussion about executing the basics. Everyone has the responsibility to keep up with the changes in our core business, because changes do come fast and furious most of the time. Just consider 30 years ago (about the time I started my career): who would have guessed that an air pack would, in the future, last for 60 minutes and become a positive pressure device? How about thermal imaging cameras (TICs), which give us the ability to look through the smoke and see the heat signatures of objects? Would anyone have believed that we would someday be drilling into a patient's bone to start an IV at a trauma scene? Or using various drugs and medications at the scene of a medical emergency to restart a heart?

The point is that we all have to be lifelong learners to be successful in this and just about any other business. I have said it often: we all must be readers of technical and other new information. There is no way that a member (career or volunteer) can get all the information needed for a lifetime during their initial training phase. Times do change. Operations do improve. Procedures are challenged and regularly revised. With so much happening around us, we have to keep our skills sharp and our processing abilities at the highest level achievable to perform a very difficult job—leading during communities' emergencies.

In Dr. Stephen R. Covey's bestselling book, *7 Habits of Highly Effective People* (1989), one of the habits is "sharpening the saw" on a regular basis. The saw is a metaphor for the skills and ability of the leader. Over time, without care and sharpening, the blade becomes dull and cannot cut. It becomes ineffective at performing its job and is left on the shelf. Covey cautions leaders to constantly improve themselves by keeping their good old saws (their skills and knowledge) sharp (up-to-date and in-practice) enough to do the job at hand, rather than allowing them to become dull and ineffective. What a great analogy for keeping leaders motivated to stay informed and focused on being capability of performing.

CRITICAL LEARNING POINTS

1. Have and maintain all of the basic national certifications for fire, hazardous materials, and rescue responses, not just EMS.
2. Determine what your department's core values are and work to improve the skills, knowledge, and abilities that are important to your department on every shift or duty day.
3. Many new technology and procedural items have been added to our profession to provide better care and services to our customers. To be effective, we must keep up with all the changes that affect our industry. All leaders must be readers and lifelong learners! It is great to be an effective writer as well, but reading is a must to consider yourself a leader of people.
4. As Stephen Covey points out in his leadership book, keep your saw (self) sharp. Every day is a training day! Most of our didactic skills will quickly erode without a comprehensive training program to keep our skills level high.

LEADERSHIP DISCUSSION QUESTIONS

1. Describe, in detail, how you go about maintaining your certifications. How often must you recertify for your various levels of attainment? How do you file and maintain the certifications that you already possess for rapid access?
2. Why is it important to keep up with new information and technology changes?
3. List three major changes you have experienced in the delivery of fire protection, three for the delivery of EMS, three for hazardous materials response, and three for technical rescue operations. Compare these changes to the way things were 20 to 30 years ago.
4. What do you think would happen to a fire department that made no major changes at all? How would the community support it? How would it operate?
5. Do you know someone who is always obtaining new and higher levels of professional certification? If so, how is this person thought of in the department? How is this skill-leveling perceived?

6. Make three columns labeled "personal training," "education," and "experience" and list your specific responses to this question: what can I do to improve myself and be better at my job?
7. Outline, in writing, a plan for your own personal improvement to maintain and raise your level of national fire and EMS certification.
8. How does item 7 fit into your overall department's comprehensive training goals?

PERSONAL LEADERSHIP PLAN FOR IMPROVEMENT

1. What behaviors do you demonstrate that are blocking you from flawlessly executing the basics of your position and job requirements?

2. What behaviors must you adopt to flawlessly execute the basics of the job within your department?

3. List at least one training program you will attend in the next year to become a better leader for your department and flawlessly execute the basics of the job.

4. List at least one leadership book you will read in the next year to become a better leader and flawlessly execute the basics of your job.

5. List the names of the members of your department you will mentor on flawlessly executing the basics of their job.

____

____

____

6. List the name of one member you will ask to be your mentor on flawlessly executing the basics of your job.

____

7. Add one leadership preparation commitment you are willing to make in the next 12 months to flawlessly execute the basics of your job.

____

____

____

PRACTICAL APPLICATION AND CASE STUDIES

CASE STUDY 1

HOW TO GAIN COMPLETE COMPLIANCE FOR SOGS

Dennis L. Rubin

Perhaps the five most important components of using a standard operating guideline driven structural fire response system are: training, implementation, follow-up, enforcement, and revision.

I am very pleased to report that I received a lot of positive reader feedback for March's column discussing the importance of using an standard operating guideline (SOG) system to handle structural fire events. I handled about a dozen requests for model structural fire SOGs for folks to use as a template to develop their own structural fire SOGs.

Well-written SOGs are just the beginning aspect of having an effective and consistent structural fire response. So it seems logical to focus on the support systems necessary for SOGs to be effective in the field.

TRAINING IS KEY

It's likely that everyone understands that training is the key to implanting SOGs and, for that matter, just about any type of program or process. However, it is remarkable to learn of so many near-miss situations and other disastrous events that involved line firefighters who didn't know or understand the related policies.

When any member is given a role to perform at an emergency incident, a major part of their functional preparation is to be trained (initial and on-going) and to understand the other related guidelines.

The best analogy about failing to train on SOGs and protocols is like a member who is instructed to wear fire-department issued dark blue uniform pants. If the member relieves himself or herself in those same dark blue pants, it is a wonderful warm feeling, but no one notices.

So, to avoid that potential mess, I am urging every department to have a comprehensive training program that addresses all policies and protocols both in initial and on-going training regimens.

ONGOING TRAINING

Some departments train (actually only conduct a brief review) all of their SOGs at the initial recruit training level. Of course, that represents only the start of the comprehensive training process, but not a best practice by any means.

It is shocking to learn how many fire departments end their training efforts at this point. If this is the case (basic training or less is provided to the membership) in your department, trouble is brewing just around the next bend.

One of the most effective ways to develop an on-going training program is to divide the policies and procedure by 12—if there are 144 policies and protocols, the number to review each month is 12.

Once you have that number set, make sure that the minimum number of policies are properly reviewed each month and in the same order by all.

If your fire department has invested in a Learning Management System (LMS) for the training process, let me say, "Congratulations!" Your LMS is designed to track required training, like SOG training, without breaking a sweat. The LMS will save time, money and be more accurate than most other department training tracking systems.

If your department does not have an LMS, do not panic. The SOG training and record keeping that the non-LMS departments use will be whatever means the department typically uses to implement, manage and track all other training activities.

IMPLEMENTATION

Here's where the rubber finally meets the road. The structural fire response guidelines have been developed. All of the members of the department have been properly trained and all of the SOG training records have been captured.

So, it is now time to spread the word that the SOGs are in use. Every member needs to know that the time has arrived for the new guideline to be implemented.

Consider using multiple sources to share the message that it is go time. I would even include a radio transmission of the department's primary radio frequency along with all of the other means that your department has to communicate to your members.

Be flexible, as the guideline is being placed in service for the first day of each of your shifts or duty crew days. Never lose sight of the implementation goal to incorporate the guideline properly and effectively. The notion to punish the way to success is just as silly as it sounds.

Expect errors as a new operational policy is being implemented. If the companies are working towards successful implementation, do not disrupt their workflow by interjecting discipline. Patience and refresher training will be key.

D.C.'S MODEL

I had been working in the District of Columbia Fire Department for about a year or so when the assistant fire chief of operations described this process. I must admit that I was a little taken aback by his description of how we would enforce the structural fire guideline.

As it turned out, the follow-up process was amazing. in holding folks accountable with the primary focus on compliance. When a working fire dispatch (WFD) is requested by the incident commander, additional resources are sent to the alarm. The commander assigns the additional responding companies to specific assignments, as needed. WFD transmission is a good indication that the fire is a significant event. The way the system worked was that after each WFD, a guideline review was conducted after the benchmark of "under control" was reached.

The review asked for all company and chief officers to report to the command post after a reasonable time for rest and rehabilitation. Once the group was assembled, the incident commander took the entire group on a walking tour of the incident.

Each company officer described his or her actions as compared to the guideline requirements. If the actions taken did not meet the guideline requirements, the officer could explain why and describe the communication process that was used to make adjustments on the fly.

About 99 percent of the time, the actions taken lined up with the guidelines or that the variation was necessary, requested, and communicated. That leaves a very small percentage of when the actions were not correct and not justified. In those cases the resolution was acknowledging the mistake and re-training at the needed level.

LIFE-THREATENING BURNS

In one near-miss case study, the officer failed to completely check the basement for fire extension. This failure allowed the fire to travel unabated to the second floor where the company was assigned. The engine company officer and one of his company members received life-threatening burns at this significant row-house fire.

To follow up this near miss, a major training effort was implemented for all companies. Further, a 45-minute video training program that reviewed this incident was developed for future use by DCFD. The hope was this fire would provide lessons that we would never forget.

I can think of only one case where failing to follow the structural fire SOG ended in discipline. The focus of the No One Goes Home program was always on outstanding performance and constant improvement. Firefighter safety was the underpinning and provided the energy to keep this a highly regarded program during my watch.

IN BRUNO'S WORDS

Fire Chief Alan Brunacini was visiting D.C. Fire Department to assist us with our command training center development. We had started the mission-critical program of certifying all command officers and acting command officers (captains) in the Blue Card Command system.

While Chief Brunacini was in town, he was able to watch the No One Goes Home program firsthand after a working house fire. Chief Bruno perhaps summed up the No One Goes Home system the best.

He said, "What an effective way to critique and review a significant response. When the officers understand that they will be evaluated immediately after each and every major incident, it becomes a significant process improvement tool. The DCFD focus is to chase excellence and not on delivering discipline. More departments should consider using this simple but highly effective process."

A Tip Of The Helmet honor goes to DCFD Chief Larry Schultz (Retired) (fig. 3–3). He envisioned, developed, and implemented No One Goes Home program to make sure that we were going where we said that we would, to protect our members, our residents and visitors, and their property.

Until next time, be safe out there.

Fig. 3–3. DCFD Assistant Chief Lawrence Shultz (retired). Chief Shultz could be my younger brother.

CASE STUDY 2

FIVE RULES TO BEING A GREAT MENTOR

Dennis L. Rubin

One of your top leadership priorities is to help develop all of the people under your command. The desire is for everyone to be successful, engaged and contributing member of your agency. This development process needs to occur regardless of the member's level (rank, seniority, and informal status) within the organization.

Further, the leader's mentoring efforts must be independent of the member's talents, skills, knowledge and abilities. The leader must focus on making everyone a better firefighter and not just the few chosen ones.

The best leaders in our business (in any type of business for that matter) always take the time and effort to help each member to improve in every way possible. The great leaders help everyone to reach for the highest level that they can attain. In fact, being a role model, mentor, and trainer is the leader's personal obligation and responsibility to always ensure the safety and survival of their members.

One of the many common traits that great leaders seem to possess is that those around them are better from their association and supervision by that specific officer. A Tip of the Spear acknowledgment goes out to former Washington Mayor Adrian Fenty.

Fenty always took the time to develop everyone in his circle of subordinates. I benefited greatly because of the time that Fenty took to help me become a better leader and person. Mayor Fenty, this tip of the fire helmet goes out to you along with my sincerest thank you, sir.

THE FIRST MENTOR

I am sure that most everyone has had the pleasure of experiencing at least a few personal advocates, mentors, or role models in their life. Of course, I am speaking about a person's parents.

From the very beginning of life, both Mom and Dad take great interest in and provide focused support for their offspring. I don't want to get carried away here, but that is a powerful example of "membership advocacy," developing people to their best regardless of their abilities.

The leadership role of mentoring and role modeling to children is a good way to think about that role in the fire service. Most folks have had a parental figure in their life and may now have children of their own.

So, the bar has been established and the training has been provided for just about everyone early in life. It is a simple matter of the fire department leader...stepping up and becoming the great mentor that lurks deep inside all of us.

Well, not so fast! Let's quickly review the rules of proper mentoring and role modeling.

FIVE RULES TO MENTOR BY

1. You must be a mentor to everyone under your command, not just the few who look like and act like you. Be a mentor for the many. Sometimes, the human condition gets in the way and we focus on the people like us. Don't fall into the trap of helping only those you like.
2. When your skills of developing people into better members are noticed outside of your department, be prepared for others to ask for your help. This is a great compliment that validates your leadership skills and abilities; honor those requests whenever possible.
3. Not everyone will be as capable of following your lead and improving at the rate and to the level that you expect. Be patient with everyone that you work or volunteer with; they just might surprise you. If you give each one your best effort, the sky just might be the limit for their improvement.
4. Allow the bright and fast learners to be your helpers in developing other folks. A great way to give other members more real responsibility is by letting them help others become better firefighters.
5. Not everyone will share your enthusiasm and motivation or moving up the departmental ladder. I've always wanted everyone to be all-consumed by their commitment to the department. However, that is not realistic, so keep your expectations of each member in the proper perspective.

Not everyone wants to be or has the skill set to be the fire chief. Be realistic in your preparation of members, but help every single one in your command to be a better member.

If you take this advice, you will be very successful in your fire-rescue service career or your volunteer vocation. Remember, a great fire department is composed of great members and officers at all skill levels and abilities.

Every single member needs to be able to flawlessly execute the basics of their job. A great department needs good firefighters, good prehospital care providers, good apparatus drivers...and good command officers not just everyone hanging on to become the chief.

One of the ranking generals of World War I remarked, "First the horses, then the men, then the officers and finally, the generals...." Take care of your people and they will take care of you. Until next time, please be safe out there.

4

Relentlessly Follow Up

Truth is stranger than fiction, but it is because Fiction is obliged to stick to possibilities; Truth isn't.

—Mark Twain

I am fortunate to be able to visit many different fire-EMS departments each year and check out how they operate. I have stopped in firehouses from Long Island (thank you Ryan Murphy and many others) to the sunny west coast of California (thank you Chief Paul Mathies and many others) and numerous other places in between. It is difficult for me not to stop in a fire station and say hello regardless of the circumstance of my travels. If the opportunity exists, I love to watch as members of a fire department apply their trade skills, knowledge, and abilities both inside (project work) and outside (response work) the organization.

Universally, I would say that the departments do a great job of relentlessly following through with emergency response tasks at hand—for instance, when the incident commander (IC) calls out on the correct tactical channel of the operational radio to reach a company for a progress report. Let's say that command is calling Division 2 while the companies are working a hoseline fighting the fire on the second floor. Once Division 2 is called and six or eight seconds pass with no response on the radio, the IC will pause another four or five seconds and call them a second time on the radio. Once that second attempt is made, the commander will wait a few more seconds before a third attempt is made to reach the targeted group of members. If the third request does not yield a response from the unit(s) in the hazard zone, a Mayday call (fig. 4–1) is transmitted by the IC. A long list of reactionary steps is initiated to locate, protect, and remove the companies that are in distress and lost. Fortunately, the root cause of this Mayday rescue example is a communications problem. The operating units are able to self-evacuate from the immediately dangerous to life or health (IDLH) environment and report their status back to command.

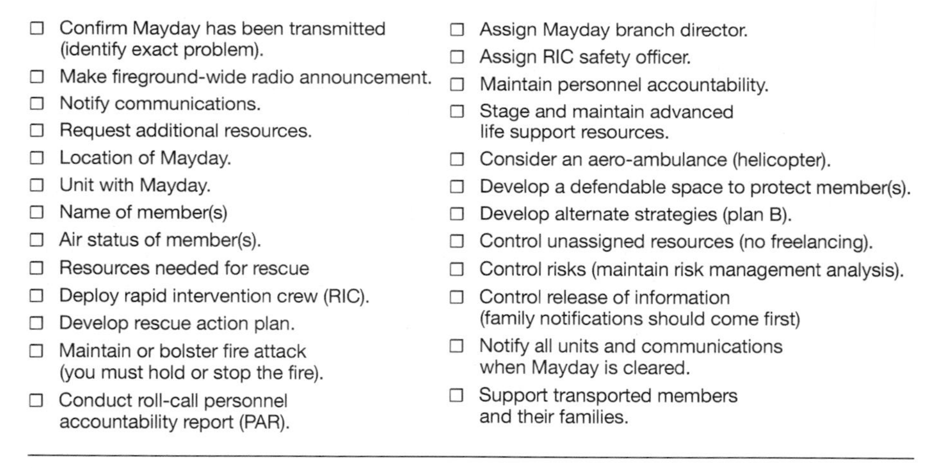

- ☐ Confirm Mayday has been transmitted (identify exact problem).
- ☐ Make fireground-wide radio announcement.
- ☐ Notify communications.
- ☐ Request additional resources.
- ☐ Location of Mayday.
- ☐ Unit with Mayday.
- ☐ Name of member(s)
- ☐ Air status of member(s).
- ☐ Resources needed for rescue
- ☐ Deploy rapid intervention crew (RIC).
- ☐ Develop rescue action plan.
- ☐ Maintain or bolster fire attack (you must hold or stop the fire).
- ☐ Conduct roll-call personnel accountability report (PAR).
- ☐ Assign Mayday branch director.
- ☐ Assign RIC safety officer.
- ☐ Maintain personnel accountability.
- ☐ Stage and maintain advanced life support resources.
- ☐ Consider an aero-ambulance (helicopter).
- ☐ Develop a defendable space to protect member(s).
- ☐ Develop alternate strategies (plan B).
- ☐ Control unassigned resources (no freelancing).
- ☐ Control risks (maintain risk management analysis).
- ☐ Control release of information (family notifications should come first)
- ☐ Notify all units and communications when Mayday is cleared.
- ☐ Support transported members and their families.

Fig. 4–1. Mayday command checklist

We are great at relentlessly following up at most emergency incidents when our members' lives are at immediate risk. However, sometimes we can get sidetracked when less motivating situations are occurring. Perhaps the routine (sometimes viewed as mundane) issues are the toughest to get properly completed in a timely manner. Follow-up is a lot more difficult for the less exciting, but still organizationally and professionally important items, such as scheduled public fire education, smoke alarm installation, or fire prevention inspection.

Never lose sight of the fact that following up does have its rewards. At a seminar on quality improvement, a fire chief mentioned that the city he worked for had received the Malcolm Baldrige National Quality Award for being one of the nation's best-managed cities. It was easy to see and hear the pride in the face and voice of this public-sector executive. After discussing the award for a few minutes, the chief joked that key among the qualifications his city possessed was the ability to return the one telephone call the evaluation team had placed to this large southwestern city's switchboard. As the Malcolm Baldrige evaluation team would tell everyone, ten calls were placed to ten major cities around country, but only two cities were willing to follow up and learn about the opportunity to be declared one of the country's best-managed cities—and his community was one of those two.

There were many, many more performance measures that were evaluated and the entire Baldrige process took over six months to complete, but the very first step to being considered was to simply follow up by returning a telephone call.

I had a similar experience when I worked in Atlanta. The Fireman's Fund Insurance Company (FFIC) had selected the greater Atlanta area to be the host region for the second phase of their nationwide kick-off of the now very popular Heritage Foundation Program. I received a telephone call from one of FFIC's vice presidents, Darryl Siry, who wanted to set up a meeting to discuss how his company would provide equipment and significant operational resources for the department and other emergency response agencies in the Atlanta metro area. I was very skeptical about agreeing to host this vague planning meeting and the "sounds too good to be true" caution lamp was lit as Mr. Siry described this event. However, he had already set up a time and a date when he and his team would arrive at my office to discuss their plan. It ended up being very lucky for Atlanta Fire-Rescue that I was available to meet with the Heritage Foundation Team with little notice and little background information. Having met with dozens of sales folks, I thought that there had to be a hook of some type, but the FFIC Heritage Foundation Program was exactly what Mr. Siry had described. The Atlanta metro area was awarded nearly $500,000 in various grants and Atlanta Fire-Rescue was the recipient of nearly a quarter of a million bucks, simply because we followed up on a request to meet.

Failing to follow up on a specific issue may have results ranging from "no harm, no foul" to disastrous. There are stories of failing to follow up that are of epic proportions. When you make a commitment inside or outside your organization, there is someone expecting you to perform the obligation on time, accurately, and professionally. When an organization fails to follow through with a commitment, no excuses can help you regain the public trust that is lost.

The DC Fire-EMS Department had a very aggressive smoke and carbon monoxide detector installation program during my watch (until the public safety committee chief cut the funding). As the fire chief, some of the worst news I could get was that we had committed to installing a detector and failed to complete the task within the timeline agreed upon with our customer. Our ability to hold the public trust, liability issues, organizational reputation, and the agency's integrity all came into question when we failed to meet our obligation. The senior officers who oversaw this program were regularly reminded of the importance of this mission and consequences of failing to meet the agreed schedule. Further, I conducted spot checks on our smoke detector hotline at least monthly. Of course, once the word got out that this program was being monitored and measured, positive results closely followed.

I would say that relentless follow-up and feedback goes directly to the organization's reputation and the reputation of the individual that is making a commitment on behalf of the agency. The follow-up process is not very enjoyable for most firefighters or perhaps anyone. It is the least exciting element of a program and/or project when we are all geared to handle the exciting stuff (fire calls and medical emergencies). However, when handled well, it will raise credibility and trust in all aspects of your agency (for that matter for your personal reputation as well).

Here is a real-life example of using relentless follow-up to achieve a very specific and important goal. April 16, 2006, was a devastating day for the Atlanta Fire-Rescue Department. That evening, an automatic fire alarm activation response was received at the Hartsfield International Airport, Fire Station #24. Fire Apparatus Operator Russell Schwantes was on duty that evening and reported to the vehicle that he was assigned to drive to start the response to this call. After a few seconds of discussion, Russell was made aware that the call was for Engine 24 and not the unit that he was assigned to, Crash 24. At that point, Russell returned to the rear of the fire station to complete his on-duty mandated physical fitness program.

Shortly after returning to his aerobic exercising, Russell experienced severe chest pains. One of the on-duty medics provided the initial care, starting with a primary survey using the manual defibrillator paddles to get a "quick EKG look." The determination was made that Russell was in need of definitive medical care and was immediately transported to the Atlanta Medical Center. Nine days later, Russell Schwantes died as a direct result of the emergency responses that he attended that day. Atlanta Fire-Rescue Department Fire Apparatus Operator Russell Schwantes's name would eventually be added to the Roll of Honor wall in Emmitsburg (MD), acknowledging his tremendous personal sacrifice to his community and to our country.

All of the detailed and lengthy paperwork was completed and submitted to the Department of Justice (DOJ), requesting the Public Safety Officer Death Benefit under the Hometown Heroes Survivors Benefits Act of 2003. This action would provide Russell's surviving family members with the more than $300,000 line-of-duty death (LODD) benefit. The DOJ held Schwantes's application in abeyance and did not take action on his LODD application. Several public hearings were held requesting that affirmative action be granted in favor of approving the LODD application and allowing Russell's surviving family members to be compensated for their loss. No such luck—the word from Washington was not very positive. The application would languish for several more years without a decision being made by the Feds.

In the meantime, Russell's widow, Athena (fig. 4-2), would swing into action to ensure that her young family was cared for in Russell's absence. Athena became the face and voice for the Hometown Heroes Survivors Benefits Act. Rather than simply waiting to be put off by the DOJ, Mrs. Schwantes was relentless in her follow up to get the agency's approval of the LODD benefit application. It was amazing and rewarding to watch her put a face and name to the many Hometown Heroes applications that the DOJ would not act upon without explanation.

Fig. 4–2. I was honored to participate at the opening of the 2008 National Fallen Firefighter Ceremony with Firefighter Russell Schwantes's widow Athena Schwantes and daughters, Congressman Steny Hoyer (MD), President George W. Bush and DC Honorary Fire Chief Hal Bruno.

Among the many ways that Athena was relentless in her follow-up was making appearances on Capitol Hill (showing up) to testify before the US Congress, speeches at the many National Fallen Firefighter Foundation (NFFF) Whistle-Stop Tour locations throughout America, and national appearances at various fire-rescue conferences every chance she got. After two years of very dedicated work, the DOJ changed their position and finally approved Russell Schwantes's application for the National Public Safety Officer Death Benefit payment. Only through Mrs. Schwantes's relentless follow-up and actions does the American Fire Service enjoy this important benefit today. All American firefighters owe her thanks for a job well done. Athena role-modeled the exact behaviors I am pointing out in this chapter. Here is a tip of my fire helmet to her!

SUMMARY AND REVIEW

One of the most important aspects of upholding the public's trust is to follow up on any commitments made to our customers. If you have agreed to conduct a fire prevention inspection or install a smoke and carbon monoxide detector, be at the correct location at the appointed time or perhaps a few minutes early ready to handle the request at hand. There will be times that emergency response work will require an in-service company to divert to an alarm. Simply make sure that the party awaiting the fire department's arrival is aware of the delay and the reason for the tardiness. Next, ask to reschedule the visit as soon as possible. The communications center, the department's administrative offices, another fire company, or even the on-duty supervisor (battalion or shift commander) who was not called away to respond to the unfolding event can make that all-important follow-up call to the waiting party.

When we follow-up on all issues, there can be significant rewards as well. Consider the Russell Schwantes case or the numerous other instances mentioned in this chapter where being persistent and following up provided benefits to those who persevered.

CRITICAL LEARNING POINTS

1. Following up at emergencies is second nature for most departments. But you must also be relentless in following up on other issues; in this way, the department's reputation will be elevated and the public trust will be upheld. Never underestimate the impact that being diligent in completing your duties has on the community. When a barrier, such as

an emergency response, is placed in the path of accomplishing these duties, simply let the waiting party know the situation and reschedule the event.

2. The possible benefits of relentlessly following up, both by completing duties and by "calling back" on opportunities, are significant.

LEADERSHIP DISCUSSION QUESTIONS

1. Describe at least two situations in which you have relentlessly followed up to fulfill a personal commitment.
2. Describe at least two situations in which your fire-rescue department has relentlessly followed up to fulfill an organizational commitment.
3. Describe and discuss an experience where you personally failed to follow up on an important issue (perhaps forgot something). How did you recover from this event? How did you feel about failing to keep your word?
4. Describe and discuss an important situation where your department failed to follow up after a commitment to do so was made. What was said about the department after the failure? How did the department recover from this event?
5. What processes can be put into place to ensure that members always relentlessly follow up on commitments? What about the department?
6. Refer to the Mayday command checklist and discuss each item's importance to the process of relentless follow-up when a member is lost, injured, trapped, or disoriented.
7. Access a Mayday case study (for example, one of the National Institute of Occupational Safety and Health [NIOSH] online reports to discuss. In your opinion, did the fire department conduct a relentless follow-up process to rescue their own?

PERSONAL LEADERSHIP PLAN FOR IMPROVEMENT

1. What behaviors do you demonstrate that are keeping you from relentlessly following up?

2. What behaviors must you adopt in order to relentlessly follow-up?

3. List at least one training program related to this topic that you will attend in the next year to become a better leader for your department.

4. List at least one leadership book related to this topic that you will read in the next year to become a better leader for your department.

5. List the names of the members in your department whom you will mentor on relentlessly following up.

PRACTICAL APPLICATION AND RELATED CASE STUDIES

CASE STUDY 1

KEEPING THE PROMISE: THE FIREMAN'S FUND REIGNITED

Dennis L. Rubin

You never know where a simple telephone call may lead. In February 2004, I received a very interesting and, at the time, somewhat vague call. As I answered the telephone with my official title and "business voice," the gentleman on the other end introduced himself. The opening few minutes of the call went something like this: "I am Darryl Siry of the Fireman's Fund Insurance Co. and I would like to set up an appointment to meet with you, Chief. We are establishing a benevolent program entitled Heritage Fund, whereby we will be making sizeable donations to fire service agencies across America."

Not wanting to miss an opportunity, but being very skeptical about the sincerity and true value of such an open-ended offer, I agreed to meet with him and one of his colleagues. Boy, what a marvelous surprise was in store for the American fire-rescue service that will continue to provide mission-critical resources for many, many years. The Fireman's Fund Heritage Program delivered big in its first year and promises even more support to our agencies in years to come.

Fireman's Fund was founded in 1863 in San Francisco with a mission to pay 10% of its profits to widows and orphans of firefighters. At that time, catastrophic fires were commonplace. In the wake of such devastating fires, there was financial loss of huge proportions. It was the era when "fire marks" were displayed on insured buildings to provide a monetary incentive for local fire companies to work a little harder to put the fires out quickly in order to limit the damage.

During these all-out firefights, it was not uncommon for members to lose life and limb while protecting lives and property. Sounds like today, but there were no controls such as the incident management system or rapid intervention teams to protect our forefathers. Once they were killed or injured, there was little or no support offered to the surviving family members. In a true effort to provide for the care and support of firefighters' families, Fireman's Fund Insurance established a trust fund (perhaps the very first one) to help with the financial burden of a line-of-duty death or injury of a firefighter.

As time marches on, change is inevitable. Fireman's Fund moved away from its original promise to help firefighters. With programs like Social Security and the Public Safety Officers Death Benefit, it was easy for the company to lose sight of its pledge to help us. To emphasize the different direction that the company had taken by the 1980s, Fireman's Fund sponsored the Major League Baseball "Relief Pitcher of the Year" award. It was quite a nice award for the American League and National League relievers who posted the best "save" records. The trophy was a silver Cairns "Senator" replica helmet atop a wooden stand; this was very nice, but of no help to America's true heroes, the fire-rescue service.

The company reorganized a few years ago, identifying and hiring an entirely new executive team. At the helm is a true friend of the fire service, Chuck Kavitsky, the CEO. The turnaround team quickly realized that the promise that the company was founded upon must be upheld. Another reason why Fireman's Fund has restated its commitment to the fire-rescue service is the background of the chief operating officer, Joe Beneducci. He is the son of a New York City firefighter. When you talk to Joe, he will quickly tell you about his dad and the days he spent growing up in the local firehouse. When I think about my friend Joe, I now associate him with the two Fireman's Fund mottoes: "Play to Win" and "Honor the Promise."

The Fireman's Fund Heritage Program was launched in the spring of 2004. Three major cities were identified to be the test sites for this effort. Of the three cities, metro Atlanta was selected as the recipient area; hence, my meeting with Darryl Siry. He asked me to identify significant additional or specialized equipment that Atlanta Fire-Rescue might need. The only difficulty with fulfilling his request was prioritizing our tremendous needs.

After a few days of discussions with our staff and member associations (IAFF Local 134 and Atlanta's Brothers Combined Inc.), the decision was made to request thermal imaging cameras (TICs). Prior to the Fireman's Fund donation, Atlanta had only two TICs divided among 34 fire stations. Not a very effective or workable arrangement to provide this mission-critical technology at emergency incidents. When Darryl asked for my "wish list," I requested a TIC for each ladder (truck) company, which would require a total of 14 cameras.

Atlanta Fire-Rescue received all of the machines that we asked for, which just about guarantees that two cameras will arrive on every structural fire scene. Two ladders are dispatched on the initial alarm and upon transmitting "working fire," a third ladder truck is sent. The initial action plan (IAP) strives to place a TIC inside the structure with the primary search team and the second camera helps to outfit the rapid intervention team. Only through the kindness of the Fireman's Fund are we able to have this quantity and quality of equipment.

There is yet a second component of the Heritage Program. Fireman's Fund agents and affiliates are committed to providing volunteer help to support fire service programs. In Atlanta's case, we asked for and have received help with the staffing of our Atlanta Smoke Alarm Program (ASAP). On the second Saturday of every month, we hit the streets installing, checking, and maintaining residential smoke detectors. Seven fire companies merge into one geographical area to provide this service. All companies remain in service during an ASAP "blitz" day. For the program to work effectively, we attempt to get at least 14 community volunteers (two per company) to help. We can always count on Fireman's Fund to provide the volunteer staff that we need. In fact, once the members of the Atlanta-based team came out for their first community-based experience, they were willing to fund the cost of the detectors for the year.

The Atlanta metro area was awarded over $500,000 for the 2004 ASAP program. This money touched 16 fire departments as well as the Georgia State Fire Marshal's Office. One of the grantees turned a portion of its funds back into the company, which was just enough to fund our ASAP Program for 2005.

To ensure that every fire department in America receives some direct benefit of its kindness, Fireman's Fund granted the International Association of Fire Chiefs (IAFC) $250,000 to develop and implement a near-miss accident-avoidance program. Chief John Tippett of the IAFC staff heads this program.

"The IAFC near-miss initiative will, without question, save firefighters' lives," he said. "The Fireman's Fund Heritage grant was just what the near-miss program needed to have a national or perhaps, a worldwide positive impact on firefighter safety."

To check on the status of the Heritage Program in your community or to learn more about it and get it started in your hometown, or click on www.FFIC.com/heritage.

5

Communicate

The art of communicating at emergency incidents is a mission-critical process—be great at it.

—Alan V. Brunacini

Without a doubt, communication is critical in every phase of emergency service work (preparation for, responding to, and recovering from). The effectiveness of our communication on the emergency scene will likely determine the success or failure of our operations. In fact, after operational tactics and strategies and incident action planning and execution, proper and effective communications is the most important process that we engage in. The lives of both public safety responders and civilians hang in the balance of being able to effective and efficiently communicate at emergencies. When communications are effective, all seems to go well and when they don't, disaster occurs just about every time. I can recall one such disastrous event that took the lives of five firefighters in Hackensack (NJ). The after-action review report specifically mentioned that "there was a lot of talking, but no communicating occurring" at a major car dealership that burned (NJ Bureau of Fire Safety 1988).

The ability to communicate extends to the firehouse and non-emergency issues as well. If the department does a poor job of communicating at emergencies, the communication process is probably just as bad in nonemergency situations. Getting the message out correctly the first time, every time, is the best goal for results-oriented communications.

There are many brilliant case studies that identify examples of great communication and an equal number of case studies that describe very poor processes. Both will be explored in this chapter, along with ways to ensure that your critical communication processes are as effective as possible. Nonemergency situations require effective communications as well. The focus of this section will be happens while we are working on the streets and in- quarters.

The $64,000 question to be considered is whether two people can have perfect communications under ideal conditions. Perfect communications can be difficult, even under ideal conditions. Communication is a complex process: a thought is developed by a person (sender) sending a message to another person (receiver). This message must travel through some sort of medium, perhaps a portable radio, to the receiver. Generally, there is some sort of interference (background noise, accent, slang words, or low volume to mention just a few issues) that degrades the quality of the message. The receiving person then has to decode the message to provide feedback or take action. When we break communication down into its smallest steps and actions, we see that it is a complex and sometimes confusing process (fig. 5–1).

Fig. 5–1. Communication is complex

When we communicate at most alarms, all the above factors must be considered and overcome for us to be effective at the mission-critical function of communication. Transmit your message in a concise, clear, calm, and commanding (projected, but not screaming) voice. When the transmit key is activated, pause for just a fraction of a second so that (hopefully) your radio will capture the frequency. This action should ensure that the entire statement is broadcast to all at the incident. Next, always use standard messages and directions that are incorporated into your department's policy and training. All operational communications must be based on the National Incident Management System (NIMS) model, following the protocols. The day of the 10 codes and other coded messages have gone the way of the horse-drawn hose wagons (although those were the good old days)! Plain talk is a necessity for all public safety response agencies, including police departments. There will be no worries about a misinterpreted or misunderstood 10-code message if plain talk is used at responses.

Perhaps the most important concept of being effective as a communicator is using the NIMS communications order model, where the receiver repeats the message to the sender briefly but accurately to ensure the understanding and reception of a complete and accurate message. For example, when command gives the order to a truck company to control the ventilation flow path on the fire floor, the truck company assigned must restate the incident commander's message to ensure that communications between the two has been successfully completed. "Copy," "Received," "OK," or any other limited acknowledgment is just a communications trap waiting to bite an officer when they least expect it. Take the time to briefly restate the core idea of the message to make sure that the direction was received and understood.

Another example would be a conversation like this:

"Main Street Command to Engine 25."

"Engine 25. Go with your message, Main Street Command."

"Engine 25. Advance a 2½-inch handline to the second-floor bedroom in quadrant C—Charlie—and attack the fire. You are assigned as Division 2 supervisor."

"Engine 25 copies. Advance a 2½-inch handline to the second-floor bedroom on quadrant C—Charlie—and attack the fire and assume Division 2 supervisor."

"Affirmative Engine 25."

The complete idea has been communicated from one person to another with a high degree of understanding.

A running complaint that I hear about using the communications order model is that there is not enough radio time or personnel to use such an elaborate process to get a simple task communicated. The reality is that you can't afford to *not* use this system, which is intended to prevent mistakes, omissions, and duplications of tasks. The last thought for this section is to remember to use the phonetic alphabet when communicating, such as "Charlie" for C as in the above example. Using the phonetic pronunciation (fig. 5–2) will greatly increase understanding and limit confusion and misunderstandings.

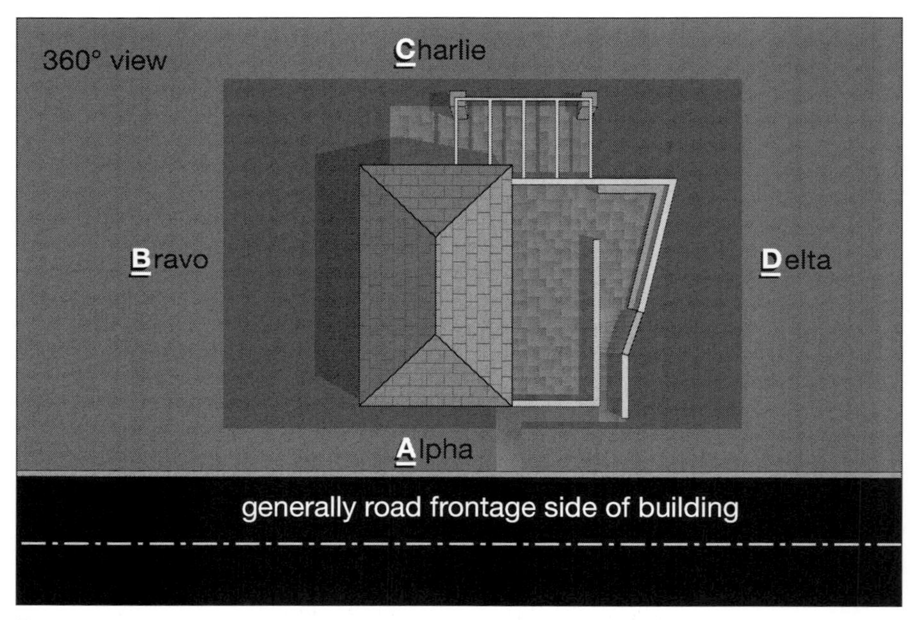

Fig. 5–2. Phonetic alphabet

There is an art to learning how to communicate at an emergency incident. There is a great deal of stress and other negative factors that must be resolved to have effective communication. Please take the time to learn more about effective communications and training on all of your department's policies and procedures as though your life depends on it, because it does!

PUT ON YOUR "BIG EARS"

One of the most commonly overlooked critical elements of effective communication is the simple practice of listening. By virtue of the two-way nature of communications, to be a great communicator is to be a great listener as well.

The old axiom, "You have two ears but one mouth—listen twice as much as you speak," always comes to mind.

On December 28, 1978, United Airlines Flight 173 was flying from Denver's Stapleton Airport to Portland International Airport. A senior pilot (Captain McBroom) was at the controls that evening and the conditions were favorable for flight. As the plane started its approach to land in Portland, there was a very loud noise followed by some control issues. The National Transportation Safety Board (NTSB) later reported that the landing gear control device broke in such a way that the landing gear was dropped out of its housing and locked into place (still suitable for landing). However, the electronic circuit that indicated the position of the landing gear was badly damaged in the free fall of the main landing gear.

Because of the anomaly of the very loud noise and the initial difficulty with controlling the aircraft, the pilot opted to abort the landing in favor of completing about a 45-minute "go-around loop" to give the cabin crew time to prepare for a potential crash landing. As the captain was discussing the plans with the first and second officers on board with him, both expressed concern about fuel quantity. There was discussion and confusion about the fuel amount and the configuration of the fuel pumps and cross-feeds, as captured on the cockpit voice recorder (CVR). The captain continued with his decision to make a loop.

About six miles outside of the airport, Captain McBroom was shocked when the jet engines began to flame out (something that happens when they're fuel starved), and United 173 crash landed. There was a significant loss of life (10 people were killed), along with 24 folks who received significant injuries. If there was any "good" news, it was that there was no ensuing fire to add to the body count, as the fuel was exhausted prior to this crash landing.

According to the final analysis of the NTSB report, the captain did not pay enough attention to the fuel load or to the advice and information being provided by the DC-8 flight crew, as it appeared that he may have been distracted with giving instructions to the flight attendant to prepare the passengers for a potential crash landing. The NTSB further indicated that the crew should have been more assertive in reporting low fuel. The lack of proper and complete communication was one of the causal factors leading to this horrendous plane crash. The landing gear was down, locked into place, and could have been effectively used to land the aircraft. The failure to understand the fuel status caused many unnecessary deaths of that day.

SUMMARY AND REVIEW

Effective and accurate communications are a major part of all aspects of public safety service. Regardless of whether the communication involves a life-safety issue or a routine fire station message, the goal is to get the information right the first time every time. Clear, concise, and direct communications are the key to properly managing all aspects of fire-rescue operations, and that must be a base belief of everyone in our industry.

One must consider that fire-rescue departments communicate between many locations (fire stations, divisions, sections, etc.), incorporating a wide range of significant issues. Other related effective communications difficulties include rotating shift work, the part-time nature of most volunteer firefighters, general attendance issues, and various types of absences from duty. Looking at the process in the broadest of terms, it is difficult to practice an effective communication process all of the time.

Another factor is that people have a preferred method of communication, one that they appreciate the most and communicate with the best. For example, I don't care for email and electronic messaging. It is difficult to see the typeface and sometimes it does not convey the entire meaning. The point is that the department needs to send out messages using as many mediums as possible. Electronic, paper, and voice messaging are examples of methods that should all be used to get the message out.

To wrap-up the leadership communications discussion, the application of a learning management system (LMS) is strongly encouraged by your author. Whenever the LMS is accessed and used to transmit a message, it ensures that the messages get to (with documented receipt) the intended receivers.

CRITICAL LEARNING POINTS

1. Perfect communication, even under ideal conditions, does not exist without effort. Leaders and followers must work hard to improve their communication skills.
2. Using plain text at response incidents is not just a good idea, it is a required component of NIMS.
3. Don't forget to use the communications order model whenever giving or receiving information at an alarm.
4. To be a good communicator, the leader must be a good listener.

5. Use as many methods (face-to-face, electronic, on paper, etc.) as you can to transmit a nonemergency message.
6. Consider using a learning management system to improve organizational communications.
7. To avoid confusion, use the phonetic alphabet. For example, side A is side Alpha.

LEADERSHIP DISCUSSION QUESTIONS

1. Have you ever attended an incident where a message was misunderstood? If so, write down all of the details.
2. List at least three ways that this misunderstanding could be avoided in the future.
3. Can you think of a company or chief officer who is a good communicator? If so, list the traits that come to mind when defining why this officer is a good communicator.
4. Diagram the steps in a two-way conversation. Diagram the steps used in the communications order model.
5. List the six steps in providing a quality brief initial report. Describe each step in detail.
6. Why is a brief initial report a critical component of a good situation size-up?
7. Why should we always use plain text at alarms?

PERSONAL LEADERSHIP PLAN FOR IMPROVEMENT

1. What behaviors do you demonstrate that are blocking you from being a good communicator?

2. What behaviors must you adopt to be a good communicator with your company and the department?

3. List at least one training program you will attend in the next year to become a better leader and communicator for your department.

4. List at least one leadership book you will read in the next year to become a better leader and a better communicator.

5. List the names of the members of your department you will mentor on the skill of communicating.

6. List the name of one member you will ask to be your mentor on the skill of communicating.

7. Add one commitment you are willing to make in the next 12 months to learn to be an effective communicator.

6

Perform Consistently

A lot of guys go, "Hey, Yog, say a Yogi-ism." I tell 'em, "I don't know any." They want me to make one up. I don't make 'em up. I don't even know when I say it. They're the truth. And it is the truth. I don't know.

—Yogi Berra, New York Yankees Baseball Hall of Famer

Perhaps the single most important trait for any organization is consistent performance in every aspect of its operational service delivery. I cannot think of a single agency that doesn't want to be consistent in how it functions and delivers services or products to its customers. The fire-rescue service (actually all of the public safety departments) is no different in seeking a long-term consistency. A comment that any fire chief hates to hear is that there are several or even many fire departments operating under the banner of his or her agency. For example, 7 battalions times 3 shifts could equal 21 small departments within a single agency. Some outfits devote a lot of time to achieving and measuring their performance to ensure that they are as consistent as possible in all that they do. Some departments place little or no emphasis on consistent performance. The agencies that ignore this critical performance element generally pay for it in poor performance. Most dramatically, at emergency incidents where firefighters perform their duties in an uncoordinated and disconnected way that is outside the scope of policy (inconsistent), they are hurt or killed.

Consistency must include the ability to perform correctly, effectively, and safely. Remember, consistently bad performance is still and will always be a bad performance. We in the fire-rescue service must strive for consistently good and even great performance because lives (including ours) depend on our services being delivered. I recommend that the department's strategic vision, mission, and goals call for consistently great performance by all members and companies all of the time. Legendary Green Bay Packers and Washington Redskins coach Vince Lombardi perhaps said it best: "We will chase perfection, and we will chase it relentlessly, knowing all the while we can never attain it. But along the way, we shall catch excellence."

The best organizations have methods to measure the results of the departments' work efforts with an emphasis on consistency and high-quality performance. This organizational goal is simple to recite and understand, but it is difficult for any agency to achieve, much less one that must work under demanding constraints (emergency nature of our work) all of the time. The department will only know when they have reached this goal if they take the time to measure their results. If you do not measure the department's processes, consistency cannot be determined or declared. Further, components and processes that are not measured cannot be improved. To ensure the department is reaching the highest level of service achievable, measure the processes. In simple terms, measure everything that is important to the customer and those items that are important to the organization (the color of a firefighter's socks does not make the measurement list, but turnout and response time should be on the list).

If the style of hose loads is based on the shift that is working that day, truly consistent performance will be difficult to impossible to reach. In some places, core operations change in measurable and visible ways from one shift to another. If there is a void in operational procedures, if training to support those procedures is not in place and sustained over time, or if an agency's policies are not enforced, the outfit will never attain consistently good performance (much less excellent performance). The elements of clear vision: written policy, initial and ongoing training, policy supervision and enforcement, and reward or discipline and plan revision, must be entrenched within the agency (fig. 6–1). The SOPs/SOGs must be described officially and supported organizationally all of the time. These elements are basic requirements to achieve the highly desirable the outcome of the consistent performance. If the leadership of the agency fails to set standards and provide the needed resources and incentives, the department is destined to varying results and varying community satisfaction levels, consistent performance will be out of the grasp for the department.

- Clear Vision—becomes the plan
- Written policies (SOP and SOGs)
- Training (initial and ongoing)
- Implementation of the plan
- Supervision and enforcement
- Reward, retrain or discipline
- Evaluate the plan and revise
- Plan revision—starts all over

Fig. 6–1. Elements of clear vision

To underscore the importance of this organizational goal, let's talk about an American institution that exudes consistency in products and performance. Many years ago, when the McDonald brothers started a hamburger stand that used production line principles, one of their core values was consistency. Think about the last time you visited a McDonald's restaurant near your home. Then think about the last time you visited a McDonald's more than 100 miles from your home. What was that experience like? I am willing to guess that the food was just like the food at your home McDonald's, that you enjoyed the same flavors and textures as at your home "Mickey D's." I would further venture to say that the restaurant was reasonably clean with the customer service that you have grown to expect from this giant corporation. I imagine the price was roughly the same, as well as the shape and size of the parking lot.

Early into this great economic venture, the management structure figured out that consistency was the way that McDonald's would be a highly competitive and profitable fast food corporation. Nearly eight decades later, the company is thriving and is a model for consistent performance. The company spends a great deal of time and energy to ensure that a customer's dining experience is a controlled and expected one. We could learn a lot from the model this fast-food outlet provides us.

The consistency journey should start with the development of clear, concise, and well-written policies. Some organizations are reluctant to put their policies in a written format due to the legal concerns. The arguments stem from the belief that if policies, procedures, or protocols are committed to paper, the agency would be held accountable for substandard performance in a court of law. In fact, there is a group of fire chiefs who think the documents need to be described as guidelines, not procedures. Regardless of the name selected for the organizational directives, they must be memorialized in print and available for frequent reference.

Next, there must be a formal way for the members of the department who will be delivering services to the public to learn and sustain the information given in the policies. Most organizations develop an initial officer training program, but the follow-up retention training is often left to chance. Systems that focus on consistency incorporate an ongoing training component that makes sure the mission-critical information is renewed and reviewed on a regular basis.

The last step to ensuring consistent performance is a follow-up and enforcement component. The ideal problem is that the chief just needs to figure out a reward system, because all of the members are following all of the policies, procedures, and protocols every time and need only positive reinforcement for their work. The reality, however, is that everyone must occasionally be guided toward compliance. A standard, transparent, and equitable follow-up system should be developed and implemented to keep the agency focused on consistent, high-quality performance (more details in chapter 12 on personal behavior).

I must include a comment here about using job aids to ensure that you get the task at hand completed correctly the first time every time. Figure 6–2 shows a Mayday job aid that provides responding company officers with a quick reference to the information command needs from them during a Mayday. The idea is that the job aid will help the officer complete a last-minute procedural check to ensure that all predetermined required elements are handled correctly when a Mayday is transmitted. Figure 6–3 shows another type of job aid.

LUNAR

L Location
U Unit
N Name
A Air supply
R Resources

Fig. 6–2. Mayday job aid

1. Confirm address
2. Incident description
3. Incident conditions
4. Request assistance (if needed)
5. Initial action plan
6. Assumption of command

Fig. 6–3. Brief initial report (BIR) job aid

We use several checklists during emergency events that keep us focused like lasers during immediately dangerous to life and health (IDLH) operations. After several rounds of "Mayday: firefighter down" training evolutions, we finalized a useful and effective Mayday checklist (refer to fig. 4–1) and added it to the resources made available to the incident commander at the command post.

A miniature operational critique should also be conducted after each working incident while still on location and before the units are allowed to return to service, to discuss whether you followed your policies at each significant event. This procedure, developed by a great operations commander, Chief Lawrence Schultz of the District of Columbia Fire Department (DCFD), has had an amazing impact on ensuring consistently great operational performance. This policy was entitled "No One Goes Home," a take-off of "Everyone Goes Home," the National Fallen Firefighter Foundation (NFFF) initiative. Once the alarm is determined to be under control, the senior operations officer gathers all company officers, division and group supervisors, branch managers, and section chiefs at the command post to discuss the events that just unfolded. The safety officer is responsible for documenting the hotwash for posterity and so it can be shared with other members not on location. Each officer is asked to describe the actions their company took and why (which will hopefully be in line with the policies) as the entire command party walks to each significant location as it is being discussed. If a policy was not followed, the reporting officer is challenged by the group leader to explain why not. There should be no attempt to add discipline to this process. This is a very powerful tool and has improved our operational consistency, in no small part because of the way it was implemented and handled (as a tool for improvement, rather than punishment).

SUMMARY AND REVIEW

Consistent performance is a mission-critical element in the day-to-day management and operation of any public safety agency. The organizational goal should always be consistent greatness, which is difficult but attainable. Many factors come into play to reach this lofty benchmark every time a fire department goes out the door to a response. Of course, as this textbook emphasizes, consistent performance must start with good leadership. The 13 rules captured in these pages is a great place to start or to improve your leadership capacity and abilities.

Further discussed was the fact that when a group of individuals is left to their own devices without a processing system or operational policy, it is unlikely that consistent performance can be obtained. A model to consider is McDonald's restaurants. A great deal of time, attention, and resources go into making sure this fast-food giant delivers a similar (if not identical) dining experience with every visit to the "Golden Arches."

Policies, training, education, experience, and job-related aids ensure that the performance of the department is outstanding at every response. It is not always necessary to reinvent the wheel to make significant and measurable organizational process. So, don't feel like your department must go down the path of improvement and consistency alone. Not at all necessary, find a great example to study and duplicate.

The third item in the Practical Application and Related Case Study section is an organizational transition plan. As a leader, your responsibility is to stay focused on consistent performance throughout your watch and to transmit the current status and consistency plan to the incoming officer. The military calls this process the change of command.

CRITICAL LEARNING POINTS

Process measurement is a critical part of reaching or improving consistency. If you do not measure your operational process, you aren't operating with enough information to understand what is happening, much less improve a service delivery process.

1. If there are apparatus and equipment changes made to suit the oncoming group at shift or duty crew change, operational consistency is not going to happen and it will look like everyone for themselves.
2. You must have a defined quality and consistency improvement process that is documented, incorporated into training, and closely followed.
3. Focus needs to be on rewarding good behaviors (something as simple as thanks for a job well done) and not on discipline. Discipline should only be used as a last resort after all other avenues for improvement have failed.

LEADERSHIP DISCUSSION QUESTIONS

1. Why is consistent performance necessary and desirable for a fire-rescue department?
2. The McDonald's model that was used by the author is a realistic and achievable example of an organization reaching consistent performance. List at least one other example and provide the details that support your position.
3. Can you identify a process your department currently measures for efficiency (e.g., response times)? What others should be added to the list of items that are measured?
4. At the beginning of a shift or duty crew assignment, should response equipment be moved about according to the preferences of the officer assuming duty? If so, why? If not, why not?
5. Why should your department use checklists at emergency incidents? How can a checklist assist with consistency? List all of the checklists that are currently used in your department. Develop a list of other processes that should or could be developed into checklists to ensure consistency at all types of the responses.

PERSONAL LEADERSHIP PLAN FOR IMPROVEMENT

1. What behaviors do you demonstrate that are blocking you from performing all your position requirements (operationally and at the station) consistently?

____

____

____

2. What behaviors must you adopt to perform consistently within your company and department?

____

____

3. List at least one training program you will attend in the next year to become a better leader and consistent performer.

4. List at least one leadership book you will read in the next year to become a better leader and consistent performer.

5. List the names of the members of your department you will mentor on consistent performance.

6. List the name of one member you will ask to be your mentor on consistent performance.

7. Add one leadership commitment you are willing to make in the next 12 months to become a consistent performer.

PRACTICAL APPLICATION AND RELATED CASE STUDIES

CASE STUDY 1

WHY NOW IS THE TIME FOR CRM IN THE FIRE SERVICE

Dennis L. Rubin

When I think about the future of our business, there is not a lot that makes me nervous or pessimistic. Having been around for several decades, I am well aware that change can seem to take forever, but still usually happens. Not trying to sound trite or aged, I can remember riding on the back step, wearing "day boots" and using a 15-minute air pack. Each of these practices has faded away in time, and looking back, I think about the tremendous safety improvements that the replacement of each of those items brought to our industry.

Some of those changes have been so complete that there are only a few of us in the active duty service that will remember them at all. To demonstrate, try this little experiment. Next shift at the coffee table, conduct a survey of "who remembers." Perhaps get the folks to show hands if: they ever drove an open-cab truck, used a filter- or demand-style mask, or engaged an aerial ladder stabilizer device on a tractor-drawn aerial turntable. Odds are that not many will...have experienced those "good old days."

CRM BACKGROUND

This section will not be about the past. In fact, it is about arguably one of the most profound changes that should be taking place in the American fire-rescue service, but as yet does not seem to be gaining any traction. So little progress is being made in this area that I feel compelled to use this space to shout for help—your help, to be exact. I know that change (and, in particular, meaningful change) takes a lot of time—I first wrote about this concept in *Firehouse Magazine* in the 1990s.

A little over 10 years ago, I had the wonderful career opportunity to attend a world-class lecture presented by an airline pilot, Captain Al Haines. As you may recall, in 1987, Haines was the pilot of United Airlines Flight 232, which crash-landed in Sioux City, IA. Although it was a high-impact crash (one in which is it unlikely that anyone would survive), Haines and crew—including the air traffic controller—managed to save 182 of the 296 souls on board that afternoon. This was a feat that had been previously unheard of in the aviation industry.

Toward the middle of Haines' incredible presentation, he spoke about the commercial aviation industry's human-error-avoidance and checks-and-balances system known as crew resource management (CRM). As Haines recounted that fateful flight for the audience, it hit me: we must adopt, modify, and fully implement CRM to improve firefighter safety and survival. To add to the importance of implementing CRM, it will greatly enhance the operational outcome, thereby, reaching our top two strategic goals of saving lives and property.

QUICK SYSTEM REVIEW

CRM uses the basic philosophy, penned by the great poet Alexander Pope, "To err is human." In other words, people are always going to make errors. (Remember the axiom, "Pencils have erasers because people make mistakes"?). Knowing that, the aviation community has built in a great set of checks and balances to ensure that human inputs (decisions) during flights are flawless and error free.

The investigation of a 1978 plane crash determined the cause to be totally due to human inputs (decisions). While en route to Portland, OR, the fuel supply of United Airlines Flight 173 was completely used up. Both the co-pilot and the second officer warned the pilot several times of the fuel situation, but based on the flight line operations of the day, the captain in charge of the DC-8 jetliner disregarded their warnings, which resulted in a fatal crash. Ten people on the plane were killed (including the second officer) and 23 more were critically injured.

The only redeeming factor that surfaced on crash day was that there was no fire associated with the crash of Flight 173 (the impact took place in a Portland suburb) because the fuel was completely used up. If ignition had occurred, it is widely believed that all aboard would have likely perished.

In the aftermath of Flight 173's crash, CRM was born. By 1980, all commercial airlines were required to place great emphasis on the avoidance of human errors by implementing CRM. As a result, the aviation industry has seen remarkable results in eliminating, avoiding, or trapping human-error issues before there are any consequences (unplanned events, i.e., crashes). Even though the captains have the final say for all flight operational decisions (same as incident commanders), they are trained to listen and use a multitude of inputs before taking any critical action, including travel-route and/or emergency decisions.

In addition, the other flight crewmembers such as the first officer, flight attendants, maintenance folks, and air traffic controllers, to name a few, are thoroughly trained to be bold and assertive to make sure that the captain gets the critical input necessary to safely operate the aircraft.

The results of full implementation of CRM have been phenomenal. Human-error accidents are now almost nonexistent in commercial aviation. Further, other organizations have picked up the CRM process and, after only a few years, have seen a great reduction of accidents and personal injuries.

As an example, one of the military branches reported a 71% reduction in accidents and injuries after CRM was integrated into its air and field operations. I have only heard glowing reports about results relating to the utilization of the CRM process.

THE CONCERN

After about 10 years of discussion, only a very few fire-rescue agencies have developed, adopted, and integrated CRM into how they do business on the streets. My personal notation is that a comprehensive CRM program for our industry would greatly reduce firefighter injuries and fatalities, as well as greatly improve the service that we deliver to our customers every day.

Researching just about any firefighter fatality report from the National Institute for Occupational Safety and Health (NIOSH) will support this argument. While referencing the material, pay special attention to the incident causes and recommendations, as well as the root causes of the accidents that led to members' deaths. The typical findings will be that most firefighter fatality scenes are riddled with human errors.

The operational error that seems to occur most often is the failure to implement (or only partially implement) the incident command system, as well as the failure to identify and activate an incident safety officer. Without question, the complete application of CRM will eliminate this and all other human input errors, if used properly. A recent seminar I attended theorized an 80% or greater reduction in performance errors would be realized as a result of the application of CRM.

As I mentioned earlier, our armed forces are using CRM with great results. Beginning with the various aviation operations and then blending into all other functional operations, the transition of application of

CRM for the military was relatively easy. The result of CRM speaks volumes and cannot be denied. The next logical step is to drive it further into each organization.

The medical field is now embracing CRM in an effort to eliminate human errors and omissions that occur each year at a rate of 1% to 1.5%. An excellent performance rate of 98.5% sounds like a great track record, unless you or a family member is part of the unfortunate group in the remaining percentage. Stories of being the wrong person under the knife are real and CRM will become a major tool for the medical field to reduce or eliminate these types of needless tragedies. Interestingly, Dr. Robert L. Helmreich, a professor of psychology at the University of Texas who is considered to be one of the CRM giants, has coauthored a book (*Culture at Work in Aviation and Medicine: National, Organization and Professional Influences*) and consults to medical care practitioners.

THE CHALLENGE

The American fire-rescue service must develop a comprehensive plan to fully implement Crew Resource Management. Further, if we are going to reduce firefighter fatalities and injuries anytime soon, we must take action on this issue quickly.

It is my hope that all of the major fire-rescue service organizations will accept this challenge and "run with the ball." Ideally, it is my desire that the National Fire Academy (NFA) will develop a comprehensive resident program as well as an abbreviated off-campus offering of CRM. Further, a substantial training component on CRM needs to be integrated into every NFA operational training course.

Next, it would be wonderful if the International Association of Fire Fighters (IAFF) and the International Association of Fire Chiefs (IAFC) would adopt and urge its members to use the national model developed by the NFA. Although the IAFC is doing an extraordinary job with the near-miss reporting program, it would make more sense to me to develop the strategy to prevent these near misses in the first place.

I will close with an excellent example that everyone should attempt to model. The Phoenix, AZ, Fire Department has developed an 80-hour "Command Officer Training" course. The content is loaded with CRM components. Having spent a few days attending the program at Phoenix's Command Training Center, I can truly say that the process is leading edge. Maybe Phoenix could help the NFA to develop a national CRM training curriculum.

CASE STUDY 2

GOVERNMENT OF THE DISTRICT OF COLUMBIA FIRE AND EMS DEPARTMENT: DEPARTMENT TRANSITION PLAN

Dennis L. Rubin

OVERVIEW

The Executive Team of the District of Columbia Fire and EMS Department developed this transition plan in an effort to create a roadmap describing future growth of the Department as we embrace positive change. This transition plan will play an important role in identifying key objectives and action items to assure progress and success in the future.

MISSION STATEMENT

The mission of the Fire and Emergency Medical Services Department (FEMS) is to promote safety and health through excellent prehospital medical care, fire suppression, hazardous-materials response, technical rescue, homeland security preparedness, and fire prevention and education in the District of Columbia.

CORE VALUES

- Respect for community
- Diversity of personnel
- Rich in tradition
- Commitment to public service
- Commitment to community service
- Commitment to customer service
- Personal integrity
- Compassion for victims of fire
- Compassion for the sick and injured
- Accountability to the public, the organization, and each other
- Courage to face adversity and danger
- Be nice to everyone

2007 ORGANIZATIONAL SWOT ANALYSIS

STRENGTHS

- Quality of personnel: knowledge, skills, and abilities (KSAs), diversity, education, motivation, desire, loyalty
- Department culture and traditions
- Leaders with strong interpersonal skills
- Flexibility and ingenuity despite less than ideal conditions
- Effectiveness despite limited space/facilities
- Stations strategically located
- Response capability
- Strong SOGs
- Command and control
- Aggressive fire ground tactics
- Operational safety
- Apparatus reserves: number and type of vehicles
- SCBA: gear is functional and appropriate
- Competitive salary and benefits
- Relationship with the mayor
- Ability to work with the government
- Wellness program

WEAKNESSES

- People (operational)
- Vacancies
- Youth of department
- Certain aspects of cultural diversity
- Variability in literacy and numeracy
- Lack of succession planning
- Office of unified communications relationship
- Lack of administrative support personnel and equipment
- Public perception and trust
- Limited ability to recruit district residents as employees

- Multiple work schedules
- Status of reserve apparatus and vehicles

OPPORTUNITIES

- Broaden recruitment efforts/targets
- Better define the district's needs
- Abate personnel apathy
- Accelerated change through EMS task force involvement
- Increased training
- Improve safety through better personal protective equipment (PPE), seatbelt usage, wellness programs, rehab, etc.
- Community partnership
- Improve customer service
- "Convenience" district council for better funding
- Improve internal communications
- Access Department of Homeland Security (DHS) funding
- Develop promotional processes
- Develop additional career paths
- Enhance professional appearance
- Merge fire and EMS cultures and benefits.
- Create department-wide uniform standard.
- Work with the government.

THREATS

- Public perception: race, economics
- Budget
- Membership overtime
- Labor structure
- Politics
- Resistance to change
- Sabotage
- Aging infrastructure
- Compliance with standards.
- Internal conflict/morale

- Turnover without succession planning/sharing knowledge

KEY DEPARTMENT ACCOMPLISHMENTS 2007-2010

- Implemented an integrated fire and EMS services delivery strategy involving paramedic engine companies (PECs), medic units and ambulances that continues to improve first response EMT and first response advanced life support (ALS) capacity to sick and injured patients.
- Implemented the National Registry EMT (NR-EMT) certification standard for EMS. More than 95% of the operational workforce is now NR-EMT certified, with the complete transition expected during FY 2012. The NR-EMT program assures that all District of Columbia–certified EMTs meet a national training standard as outlined by the federal Department of Transportation (DOT), the federal agency responsible for national EMS oversight.
- Implemented a quality assessment and improvement program that monitors critical events associated with EMS calls. Using the electronic patient care reporting (ePCR) documentation system, this program allows the department's quality management team to more quickly identify compliance issues and work with employees to improve patient outcomes.
- Implemented medical quality monitoring of ST-elevation myocardial infarction (STEMI), pulmonary edema/congestive heart failure (CHF), asthma and cardiac arrest outcomes using the consortium of US and International Major Metropolitan Municipalities EMS Medical Directors (Eagles) evidence-based performance measures and the Cardiac Arrest to Enhance Survival Registry (CARES).
- Implemented a comprehensive water supply delivery program, placing in service a dedicated and immediately available water supply engine company equipped with 4 in. diameter hose within each battalion. These fire trucks deliver improved water supply capacity to incident scenes at greater quantity and pressure. The department is in the process of replacing outdated 4 in. hose with new 5 in. supply hose on all fire trucks.
- Implemented new water supply procedures and developed an in-depth water supply training program to better prepare firefighters for managing water supply issues during structural firefighting.

- Implemented an EMS patient customer satisfaction survey now mailed to all patient customers transported by ambulance. The more than 6,000 returned customer surveys indicate an overall satisfaction rate of 94% in CY 2008, 95% in CY 2009 and 97%, to date, during CY 2010.
- Implemented a completely revised employee discipline system based on fairness and consistency, along with developing better data collection and tracking systems to bring employees into compliance with Department rules and regulations.
- Implemented a streamlined trial board adjudication system for speedier resolution of serious discipline cases and developed a process to send less serious violations to chief officer hearings, thereby reducing the financial burden of more costly trial boards.
- Implemented the mobile data computer terminal program, providing all fire trucks, ambulances, rescue squads, and command vehicles with mobile computers to receive critical incident information. This system allows for quicker response times by providing the latest GIS mapping features including incident routing instructions and can account for individual unit performance by time stamping incident events.
- Implemented a "buff, scrub, and greening" program to evaluate all facilities operated by the department for priority repairs including the cleaning and painting of all hard surfaces, replacement of high use electrical lights with lower use lights, motion switches that turn lights off and on when motion/nonmotion is detected and the replacement of inefficient windows and doors with higher rated energy efficient windows and doors.
- Implemented a renovation and replacement program to modernize all facilities operated by the department. To date, this includes completed renovations of E-9, E- 25 and E-17. E-10 has reached the 70% completion mark. E-14, E-28, E-29, E-27 are in preparation phase with the renovation of E-29 scheduled to begin within six months. Renovation projects include a program to locate a temporary fire station in the immediate alarm area so as not to delay emergency response.
- Identified two fire stations for relocation to maximize service delivery including E-22 from Georgia and Missouri NW to Georgia and Aspen Street NW and E-26 from the 1300 block of Rhode Island Ave. NE to a location in the area of Rhode Island and South Dakota Avenue NE.

- Implemented the first EMS medical protocol revision process since 2002. Revised protocols effective March 15, 2010, include the use of new out-of-hospital medications, medical equipment, and treatment procedures.
- Promoted the second ever African American woman battalion chief and appointed her as the first woman deputy fire marshal in the history of the District of Columbia.
- Successfully planned and managed fire and EMS protection services for the single biggest one-day special event in District of Columbia history—the 2009 presidential inauguration.
- Successfully placed a fire intelligence analyst in the Washington Regional Threat and Analysis Center, allowing for fire and EMS professionals to directly share real-time critical threat and analysis services from an intelligence center for the first time.
- Implemented a pilot program to test the use of medium duty chassis for ambulances to evaluate the cost savings benefits of heavier chassis.
- Implemented a PPE replacement program to meet the recommended replacement schedule of the NFPA and PPE manufacturers.
- Implemented a high-risk business fire safety inspection program targeting restaurants, bars, and nightclubs during peak business hours to verify compliance with fire and life safety laws. In 2009, the Office of the Fire Marshall conducted more than 800 high-risk business inspections.
- Implemented a hospital transport management plan to better and more efficiently direct patients to hospitals. This plan has significantly impacted the closure and diversion status of hospitals, resulting in an 86% reduction enclosure and diverted hours and the elimination of hospital emergency rooms closed to ambulance transport by the end of 2009.
- Implemented an internal affairs office to coordinate the handling of cases involving alleged crimes and serious misconduct, which has increased public trust of the fire and EMS department workforce as well as increasing integrity of department employees.
- Implemented professional responsibility standards and created personal responsibility pledge forms that are now being adopted by peer jurisdictions. Implemented an electronic patient care reporting (ePCR) system that transitioned more than 100,000 paper patient care

reports to electronic records. More than 99% of patient care reports are now completed using computers.

- Implemented the Street Calls program for addressing the needs of at-risk high volume EMS system callers. Using physician assistants, paramedics, and EMTs, this program delivers medical and social assistance resources to identified or referred EMS patients. This resulted in a 75% use reduction of 9-1-1 and EMS ambulance transports for the original group of Street Calls patients tracked by the program.
- Implemented a number of community outreach programs including the Smoke Alarm Utilization and Verification (SAVU) program, blood pressure tracking and disease prevention clinics, childhood fire safety education programs, free child car seats and installation assistance and community CPR education. Together, these programs reduce the risk of fire and emergency medical problems for district residents, improving public safety.
- Implemented an NFPA fire safety compliance work uniform requirement, eliminating the long practice of using polyester work uniforms, which added to the significance of burns sustained during firefighting operations.
- Completely revised the department diversity training program and increased mandatory training hours.
- Implemented a first ever department-wide domestic violence training program for all employees.
- With the cooperation and assistance of the Water and Sewer Authority (WASA), implemented a customer services unit (CSU) concept tasked with monitoring fire hydrant infrastructure status in the District of Columbia. In 2009, these six units flush tested 25,000 fire hydrants, updating a WASA Google Earth application tracking hydrant location and service status.
- Revised pumping apparatus specifications to address the needs of an aging water supply infrastructure, including the implementation of 5 in. water supply hoses. Pursued an aggressive legislative agenda for improving public safety, including passing a fire-safe cigarettes statute along with recommending passage of legislation to ban fireworks, require residential fire sprinkler systems, and better regulate private fire hydrants not owned by government interests.
- Implemented the newest edits and updates to the department's SOGs.

- Improved employee safety and supervision at emergency incidents by adding an additional battalion chief on all box alarms.
- Improved employee safety by inserting the rapid intervention team in the initial response assignment (fifth-due engine) as opposed to past practice of waiting for working fire dispatch resources to arrive later in the incident.
- Reinstated the Back-to-Basics program focusing on delivering high quality training that emphasizes and supports critical basic fire/rescue skills.
- Completed firefighter self-rescue, firefighter rescue and engine company operations training for operational firefighter employees.
- Implemented a Fire Officer III career training program to improve chief officer credentialing and development.
- Procured grant assistance funding for a new foam unit, a command training center, a driving simulator, a medical patient simulator, WMD training, smoke alarms, chemical, biological, radiological, and nuclear defense (CBRNE) detection equipment and USAR equipment.
- Implemented the TeleStaff personnel management application. When fully functioning by December 2010, this system will for the first time allow complete and instant verification of employee assignment, status and overtime eligibility. This application has the potential to reduce overtime pressures by better management of scheduling demands before they occur.
- Implemented a random drug screening policy in compliance with District law.
- Implemented a comprehensive employee wellness program which includes adopting stress-testing guidelines and providing a rehabilitation program to successfully return people to work.
- Implemented new policies and guidelines regarding fire investigations enforcement and evidence collection that meet or exceed industry standards.
- Implemented a new working schedule for fleet management to maximize the efficiency of the fleet apparatus mechanic workforce.
- Implemented an aggressive preventative maintenance plan for all emergency apparatus to decrease out-of-service fleet time.
- Sponsored department employees on the NFPA 1901 committee to ensure that metro size department were properly represented on

the committee that sets the National Recommendations for Fire Apparatus.

- Implemented a redesigned ordering system for fire station supplies to assure all facilities have the necessary tools available to complete their mission.
- Implemented the use of the SAMs ordering and inventory control system to ensure stock levels are maintained.

KEY DEPARTMENT ISSUES

OCTOBER, 2010

CUSTOMERS AND COMMUNITY

PRIORITY 1

OBJECTIVE

CSU Program status.

Action Items

- CSUs removed from service on October 8, 2010.
- Personnel returned to Operations Division.
- DC Water to complete inspection and testing requirements per terms of new MOU with the OCA. Department to monitor testing program.

Notes

The Potential for Homeland Security funding of CSU program per the OCA, if available.

Point of Contact

AFC Schultz

lawrence.schultz@dc.gov

202-673-3320

PRIORITY 2

OBJECTIVE

ISO Certification report.

Action Items

- Develop Gap Analysis plan to identify issues for correction.
- Based on the plan, identify path forward.
- Implement correction strategy.
- Schedule reevaluation.

Notes

The District obtained an ISO rating of Class II. The Department should aggressively implement required changes to reach Class I certification.

Point of Contact

AFC Schultz

lawrence.schultz@dc.gov

202-673-3320

PRIORITY 3

OBJECTIVE

DC Water MOU.

Action Items

- Finalize MOU language with DC Water through OCA.
- When implemented for FY 2011, follow terms of the new agreement.

Notes

Current MOU with DC Water expired in October. New MOU removes Department from inspection and flush testing responsibilities, other than as indicated. Department objected to any new MOU language lessening the inspection and testing requirements currently provided by CSU program.

Point of Contact

AFC Schultz

lawrence.schultz@dc.gov

202-673-3320

CUSTOMERS AND COMMUNITY (CONTINUED)

PRIORITY 4

OBJECTIVE

EMS Task Force items.

Action Items

- Complete unification of workforce under Unified All- hazards model.
- Complete Firefighter and Hazmat certification for approximately 150 remaining single-role EMS employees.
- Complete NR-EMTB certification for approximately 70 remaining non-EMT certified firefighters.

Notes

District Council actions, Labor organization objections and overtime funding issues have prevented Department action on the items above.

Point of Contact

AFC Sa'adah

rafael.sa'adah@dc.gov

202-673-3320

PRIORITY 5

OBJECTIVE

EMS path forward.

Action Items

- Controlled Substance to Implementation (FY 2011).
- Continue bridge evidence-based response time/ dispatch protocol revision process improves efficiency and include the effectiveness of resource to track deployment.
- Increase the number of UDC Firefighter/NREMT-P's from 164 to 350.
- Continue implementation of Paramedic Engine Company program.
- Expand Street Calls and other demand reduction initiatives.
- Conduct internal EMT-B EMT-PEMT-I to EM training program T-P and program.
- Expand UDC Fire Science Degree program to EMS Management
- Establish paramedic training program partnership with or another local university and prepare District residents others to enter the workforce.

Point of Contact

AFC Sa'adah

rafael.sa'adah@dc.gov

202-673-3320

PERSONNEL AND POLICIES

PRIORITY 1

OBJECTIVE

Complete FY 2011 hiring plan objectives.

Action Items

- Hire up to 80 firefighter recruits with a start date of October 12, 2010.
- Advertise and hire up to 45 administrative positions by December 2010.

Notes

As of the date of this document, 52 Firefighter/ EMT candidates and 23 Paramedic/ Firefighter candidates have accepted offers of employment and will begin training on October 12, 2010.

Point of Contact

AFC Jeffery

alfred.jeffery@dc.gov

202-673-3320

PRIORITY 2

OBJECTIVE

Complete Medical Director(s) hiring.

Action Items

- Hire Medical Director.
- Hire Assistant Medical Director.

Notes

The Medical Director is a Mayoral appointment requiring confirmation by City Council. One round of interviews was completed during July 2010, and a highly-qualified candidate did not accept the Department's offer of employment.

Point of Contact

AFC Sa'adah

rafael.sa'adah@dc.gov

202-673-3320

PRIORITY 3

OBJECTIVE

Continue overtime management priorities.

Action Items

- Overtime management in FY 2011 entirely dependent on October 2010 hiring.
- Staffing of units requires daily oversight to stay on overtime plan.
- BSA overtime rules in FY 2011 require daily and pay period by pay period oversight.

Notes

Unless FY 2011 hiring plan is completed, overtime for the fiscal year will remain excessive and only increase again FY 2012.

Point of Contact

AFC Schultz

lawrence.schultz@dc.gov

202-673-3320

PERSONNEL AND POLICIES (CONTINUED)

PRIORITY 4	PRIORITY 5	PRIORITY 6
OBJECTIVE **Promote legislative agenda priorities.** **Action Items** • Review 2009 legislative agenda submitted to EOM and OPLA. • Update priorities as required. Check agenda items under consideration by PSJ or other Council Committees. • Check agenda items under consideration by OPLA, EOM, and other agencies. **Notes** Fireworks legislation passed, but in modified form with no ban. Nursing home legislation active under PSJ Committee. Placard legislation active under OPLA. **Point of Contact** GC Marceline Alexander marceline.alexander@dc.gov 202-673-3320	OBJECTIVE **Finalize employee background checking process.** **Action Items** • Awaiting final decision from DCHR on Department authority. • Background checks will begin with single role employees, then the remainder of employees. **Notes** The Department has been waiting for more than one (1) year for a final determination from DCHR concerning agency authority to conduct background checks. **Point of Contact** AFC Schultz lawrence.schultz@dc.gov 202-673-3320	OBJECTIVE **Resolve employee grievance issues.** **Action Items** • Local 3721 issues include random drug testing, physical testing, and workers comp monitoring through the Police and Fire Clinic (PFC). • Local 36 issues include the payment of back overtime which the Local won at arbitration. OAG appealed to the District Appeals Court on behalf of District Government. **Notes** The issues listed are the primary cases involving each Local. Other grievances have also been filed. **Point of Contact** AFC Lee brian.lee@dc.gov 202-673-3320

APPARATUS & EQUIPMENT

PRIORITY 1

OBJECTIVE

Complete FY 2010 apparatus purchases.

Action Items

- Six (6) engines ordered from Pierce. Pre-construction conference completed. Apparatus in the engineering stage, delivery expected June to July, 2011.
- Two (2) aerial ladders (rehabilitation) ordered from Seagrave. Apparatus in construction stage, delivery expected January to February 2011. A "swing ladder" (aerial ladder and hydraulics only) is part of this order. This will allow immediate replacement of any Seagrave aerial ladder in the event of damage or major repairs, significantly reducing out-of-service time.
- Two (2) aerial ladders (new) ordered from Seagrave. Apparatus in the construction stage, delivery expected late 2010 to early 2011.
- Sixteen (16) Ford E-450 ambulances ordered from Horton. Vehicles in the delivery phase, last unit expected by November 1, 2010.
- Two (2) International medium-duty chassis ambulances (rechassis) ordered from Horton. Vehicles in the construction phase, delivery expected in December, 2010.
- Various support vehicles, including Fire and EMS Battalion vehicles, on order from light-duty vehicle manufacturers. Vehicles in the construction phase with delivery expected between October and December, 2010.

Notes

Ford no longer manufacturing E-450 diesel ambulances. Medium duty chassis replacement now required.

Point of Contact

AFC Jeffery

Alfred.jeffery@dc.gov

202-673-3320

PRIORITY 2

OBJECTIVE

Begin FY 2011 apparatus purchases.

Action Items

- Bid package preparation for five (5) to six (6) engines, depending on cost.
- Bid package preparation for one (1) tower ladder.
- Bid package preparation for eighteen (18) to twenty (20) ambulances, depending on cost.

Notes

Apparatus specifications now being finalized for bid packages. Bid packages will be forwarded to OCP for action.

Point of Contact

AFC Jeffery

Alfred.jeffery@dc.gov

202-673-3320

APPARATUS & EQUIPMENT (CONTINUED)

PRIORITY 3

OBJECTIVE

Improve fleet management controls and reliability.

Action Items

- Hire certified emergency apparatus mechanics to fill vacancies and decrease overtime.
- Continue mechanic certification training program.
- Continue integration of technology to improve controls, scheduling and part supply management.

Notes

Fleet management, maintenance, and repair continue to be key areas of Department concern for reliable service delivery.

Point of Contact

AFC Jeffery

alfred.jeffery@dc.gov

202-673-3320

PRIORITY 4

OBJECTIVE

Continue with testing and certification procedures.

Action Items

- Finalize pump testing schedule for CY 2011.
- Finalize hose testing schedule for CY 2011.
- Finalize ladder testing schedule for CY 2011.

Notes

CY 2010 testing for pumps, hoses and ladders will be completed by December, 2010.

Point of Contact

AFC Jeffery

alfred.jeffery@dc.gov

202-673-3320

FACILITIES

PRIORITY 1

OBJECTIVE

Continue renovation/ replacement of fire stations.

Action Items

Replace E-26 (Brentwood), now located at 1340 Rhode Island Ave., NE. Department is working with DRES to identify replacement property near Rhode Island and South Dakota Avenues, NE. Replace E-22 (Brightwood), now located at 5760 Georgia Ave., NW. Department is working with the Deputy Mayor of Economic Development to identify replacement property near Bladensburg Rd. and Aspen St., NW, on the Walter Reed Hospital campus, scheduled for closure by BRAC in 2014.	Complete renovation of 10 (Trinidad), located at 1342 Florida Ave., NE. Project now more than 90% complete. Begin renovation of E-29, located at 4811 MacArthur Blvd., NW. Temporary fire station to house E-29, T-5 and A-29 now under construction at Loughboro Rd. and MacArthur Blvd., NW. Units will be transferred when complete. Estimated completion for E-29 is summer of 2012.	Complete renovation design for E-14 (Fort Totten), located at 4801 North Capital Ave., NE. Complete renovation design for E-15 (Anacostia), located at 2101 14th St., SE. Complete renovation design for E-27 (Deanwood), located at 4201 Minnesota Ave., NE. Complete renovation design for E-13 (River Front), located at 450 6th St., SW.	Complete renovation design for E-21 (Adams Morgan), located at 1763 Lanier Pl., NW. Complete renovation design for E-23 (Foggy Bottom), located at 2119 G St., NW. Complete renovation design for the Training Academy, located 4600 Shepherd Parkway, SW.

OBJECTIVE

Continue renovation/ replacement of fire stations.

Action Items

- Finalize E-1 (West End) replacement plan, currently in development with the West End project plan along with East Bank Properties.

Notes

E-22 and E-26 relocation planning received significant community stakeholder comment, not all of which was positive. The use of the Walter Reed campus for E-22 continues to be an issue.

Point of Contact

AFC Jeffery

alfred.jeffery@dc.gov

202-673-3320

FACILITIES (CONTINUED)

PRIORITY 2

OBJECTIVE

Replace and relocate fleet maintenance facility.

Action Items

- Identify a location for a replacement fleet maintenance facility, now located at 1103 Half St., SW.

Notes

Department is working with DRES to identify replacement property, potentially on the New York Ave., NE corridor. Project is not funded.

Point of Contact

AFC Jeffery

alfred.jeffery@dc.gov

202-673-3320

PRIORITY 3

OBJECTIVE

Relocate fire HQ from Grimke school.

Action Items

- Identify location for a replacement Department Headquarters building, now temporarily located at the Grimke School, Vermont Ave., NW.

Notes

Department is working with DRES to identify replacement property, potentially the Kelly Miller School on 11th St., NW, other property near 5th and E St., SW or even mixed use development on the Walter Reed campus. The African American Civil War Monument Museum will begin to occupy Grimke School in 2011.

Point of Contact

AFC Jeffery

alfred.jeffery@dc.gov

202-673-3320

PRIORITY 4

OBJECTIVE

Determine command training center location.

Action Items

- Confirm and extend temporary location, now at the Kelly Miller School, 11th St., NW.

Notes

Department is planning to move the facility to Training Academy after renovation of academy is completed. Academy renovation is in design stages, construction planned for 2014.

Point of Contact

AFC Jeffery

alfred.jeffery@dc.gov

202-673-3320

PRIORITY 5

OBJECTIVE

Determine status of P.R. Harris school.

Action Items

- Finalize lease/use arrangements with University of the District of Columbia (UDC).

Notes

Department is working with DRES to secure use of facility. Department invested $1.7M to renovate classrooms and gym for EMS training and CPAT. District Council action provided UDC unrestricted use of facility beginning in 2010.

Point of Contact

AFC Jeffery

alfred.jeffery@dc.gov

202-673-3320

TECHNOLOGY

PRIORITY 1

OBJECTIVE

Complete ePCR system upgrades and purchases.

Action Items

- Finish writing draft technical specification.
- Incorporate requirements into ambulance billing RFP (see Priority 4 objective).

Notes

Complete replacement of ePCR hardware and upgrade of Safety Pad software to new version. Purchases leveraged from ambulance billing revenue as part of ambulance billing vendor contract.

Point of Contact

AFC Schultz

lawrence.schultz@dc.gov

202-673-3320

PRIORITY 2

OBJECTIVE

Complete fire RMS system implementation.

Action Items

- Implement oversight and accountability of data entry requirements.
- Roll out mobile client application.
- Develop reporting requirements.
- Begin data integration with other applications and data reporting systems.

Notes

RMS system went live on October 1, 2010. Department wide training will continue through the remainder of the year. This system is foundational for integration of all data-sources.

Point of Contact

AFC Schultz
lawrence.schultz@dc.gov / 202-673-3320

PRIORITY 3

OBJECTIVE

Complete TeleStaff deployment.

Action Items

- Implementation of BSA overtime rules.
- Implementation of monthly reporting requirements.
- Integrate with PeopleSoft.
- Annual updates of personnel contact information.

Notes

Telestaff now used daily. Reporting requirements need identification and implementation. Full integration with PeopleSoft will improve available information and still needs to be accomplished.

Point of Contact

AFC Schultz
lawrence.schultz@dc.gov / 202-673-3320

PRIORITY 4

OBJECTIVE

Complete ambulance billing RFP.

Action Items

- Finish writing draft technical specification.
- Submit to OCP.
- Evaluate proposals.
- Negotiate with selected vendor.
- Submit to Council for approval.

Notes

Current contract expires 5-2-10. ePCR system upgrades and purchases included in RFP. Contract represents $21M revenue stream to District.

Point of Contact

AFC Schultz
lawrence.schultz@dc.gov / 202-673-3320

7

Tell the Truth, Always

The object is to win fairly, squarely, by the rules—but to win.

—Coach Vince Lombardi, Washington Redskins, 1969–70

The next leadership rule is perhaps the simplest to understand and remember. You have heard this rule many times and from many different folks throughout your life: tell the entire truth the first time and every time you are asked. Therefore it makes sense to talk about truthfulness in *It's Always about Leadership!*

Employees do sometimes withhold information, mislead, or otherwise behave deceptively in the workplace. When a firefighter or paramedic is not truthful, that act is quite destructive. We are the people most Americans trust with their lives, loved ones, and property. In most cases, once it is determined that a person lacks veracity (honesty), that person's ability to perform public safety services is greatly compromised or even made completely ineffective. No one will trust or rely on a person who has been caught in a lie or otherwise failed to tell the complete truth.

To capture and maintain the public's trust is one of the most important functions and capabilities of any public safety organization. To be successful at our jobs, we need the public to have a high degree of trust in every aspect of our abilities and our interactions with them. I offer that a firefighter has a position of greater trust than any other element of government. Firefighters and paramedics have more access, in most cases, than police officers. I have only heard of a few search warrants being needed to gain building entry and access for fire investigation purposes over my 30-year career. We are generally invited into someone's home or business without hesitation or a second thought. The expectation is that we will be resolving an emergent crisis and leaving once the situation is satisfactory resolved. These actions (rapid response, unrestricted facilities access, problem resolution, and incident termination) are exactly what occurs in the vast majority of all cases the fire-rescue department responds to. If we are to continue to have this relationship with the citizens we serve, we must earn and maintain their trust.

In North America, firefighters from coast to coast enter homes without search warrants to handle a wide variety of fire, hazmat, rescue, and medical emergencies. This can only happen if the public you serve has faith and trust that you will resolve their emergency, which will only happen if you fulfill these two overarching standards: your agency should always strive to be a *highly trusted* and *high performing* organization.

One of the best examples I can provide is when a paramedic is called to treat a significant trauma patient in the prehospital setting. In this scenario, the victim has life-threatening wounds that demand rapid and effective intervention, along with airway and circulation support, if they are to survive this ordeal. The paramedic swings into action to stop the profuse bleeding and provides support for the injured person's airway while ensuring there are adequate respirations. Once replacement fluids are started and the patient stabilizes somewhat, the field EMS provider cuts away clothing from the unconscious and unresponsive patient to continue the required care. Many times, people in these traumatic situations are not accompanied (no guardian or family member nearby), so responsibility for their fate (health, safety, and dignity) is in the paramedic's hands.

Over the course of my career, I have never been asked why a paramedic had to strip the clothing from a patient with the aforementioned injuries. In fact, if the individual survives their injuries, it is not out of the ordinary for the family to stop by the station for a visit and bring a token (homemade cookies, for example) of their appreciation for our high-performance services. We can only do our work effectively because the public places their trust in us to do no harm and solve their urgent problems. Having and maintaining the public trust is of paramount importance.

One way we can ensure that we have the faith, trust, and support of the public is to simply be truthful with them all of the time. During a speech given at the White House in January 2009, President Barack Obama referenced a quotation by Justice Louis Brandeis, saying that sunlight is the best disinfectant. This means that any action taken by a public person or agency should be done in such a way that everyone can see every element of the process and there are no surprises for anyone, especially the public. Once a department's integrity trust is called into question, it is difficult for that department to reclaim the trust it once enjoyed.

I would also mention President Ronald Reagan's favorite Russian proverb: *Doveryai, no proveryai*, meaning "Trust, but verify."

A major component of this process is getting all of your members to back the core value of telling the whole truth the first time. Everyone makes mistakes from time to time. That is part of human nature. The train jumps the tracks and the trust in the department slips when you try to cover up a mistake. Nothing could hurt your organization's reputation and standing in the community more than willfully misleading the public you are sworn to serve and protect. Being deceitful about emergency response times, numbers of available apparatus at any given time, personal training certifications, and members' criminal backgrounds are all examples of things that have been tried in the past. Most cases become a public relations disaster for the agency. These are all instances where deceptive practices were used to hide negative issues, rather than actual mistakes made by the department, but imagine how much worse the repercussions would be for misleading the public about actual mistakes or illegal activity.

When people make mistakes, the best course of action is to acknowledge the mistake, investigate why it happened, and put controls into effect that will eliminate or circumvent the area of the mistake in the future. Provide the community with detailed information about each step of this process (let the sunlight in).

If you don't acknowledge that a mistake was made, there is very little hope that corrective and preventative action can be implemented. Any action that tries to minimize or redirect the blame will be clear to all and seen as a hollow attempt to resolve the issue. Deliberately misleading or misguiding the public will come with great organizational danger and consequences. When the truth is revealed (and make no mistake, the truth *will* be revealed), the impact is debilitating because along with the inefficiency (mistake), the organization now has a reputation of being untruthful.

With today's communications and investigation techniques, it is very unlikely that a significant issue will be kept secret for long. Take your medicine on the front end of a problem or issue by being honest about it. It is amazing how effective you will become at protecting the public trust your agency needs, and the issue will get smaller over time instead of becoming larger in the dark. People can forgive mistakes; by and large, they will not tolerate liars.

Sometimes, the issue at hand is a simple personnel-related problem, such as arriving late to work. Almost every firefighter can relate to this situation and can likely describe some personal event that caused them to miss the start of line-up (shift change). No one likes to be in trouble or embarrassed in the workplace, so it is understandable that a person may think that manufacturing a story to save face is a good alternative to a bad situation. However, don't take that path! You will regret it in the long run and risk a much more severe penalty when the truth surfaces. The truth always seems to come out eventually and lies are very difficult to remember. In most outfits, coming to work late a time or two means a written reprimand or short-term suspension—not a pleasant experience, no doubt, but survivable. The best advice I can give you is to "firefighter up" by admitting the mistake and taking the consequences you have earned. And, of course, don't be late to work anymore (change the poor behavior)!

On many occasions, folks have tried to game the system and lie their way out of punishment or consequences. The end result typically is a harsher punishment or more severe consequences (fig. 7–1). Some agencies have rules that say lying, lacking candor, or misleading an investigation are grounds for termination.

The ability to simply tell the truth should be instilled into all of us. If your practice is to always follow this rule, your life will be a lot simpler, and your organization will be able to earn and maintain the public's trust in everything that it does. Once a person goes down the path of deceit, it is tough to recover and makes a bad situation much worse.

Fig. 7–1. Various news headlines you want to avoid

SUMMARY AND REVIEW

Without the trust and belief of the general public, the fire department cannot be effective in the community. Public trust is critical to having the fiscal support necessary for funding a fire department. Likewise, public trust is critical when delivering emergency medical patient care to someone that is in need. This trust is a measurement of the success or failure of the department.

Not often does a fire department lose the support and trust of its communities. However, there are a few case studies from our public safety partners that we can learn from. Once such case of a major erosion of trust is the situation that occurred in Ferguson, Missouri, in 2014 (and truly, is still occurring).

On August 9, 2014, Michael Brown was recorded stealing a box of cigars and shoving a clerk at a local convenience store. Officer Darren Wilson responded to the store's initial 911 request for assistance. Shortly after Officer Wilson's arrival at the convenience store, he confronted Michael Brown and a friend as they were walking down the middle of the street. The officer asked the two to move off the street and continued driving, but then reversed and pulled up beside Brown and his friend. A struggle ensued in which the officer's gun was fired twice, with one shot hitting Brown in the hand. Brown and his friend then ran, and Wilson ran after them. At some point during the chase, Wilson and Brown were facing each other and Wilson shot Brown six times, with a shot through the top of his head likely the fatal one.

Brown was 18 years old and was unarmed. Brown was an African American while Wilson is white. Among the many fall-outs from this case was the loss of public's trust in the Ferguson Police Department, which has been expressed through many protests and riots in the years since the incident. The following is a timeline of the incident and the actions taken in the aftermath.

August 9, 2014: Unarmed teenager Michael Brown is shot by Officer Darren Wilson and dies. His body is in the street for four hours. This event sparks several days of civil unrest.

November 24, 2014: Grand jury announces decision not to indict Wilson. Protests, some violent, break out in Ferguson and major cities across the nation.

November 29, 2014: Wilson resigns with no severance pay. His lawyer says he "will never be a police officer again" (NBC News 2014).

March 4, 2015: The US Department of Justice releases the results of an investigation into the Ferguson police force that concludes that these officers regularly violated citizens' constitutional rights with racially discriminatory practices.

March 9, 2015: City Manager John Shaw resigns.

March 10, 2015: Police Chief Thomas Jackson resigns with severance pay for a year.

The local outcry was for the mayor of the city to resign from his post as well, but he was reelected in 2017. Finally, there is much discussion about dissolving the Ferguson Police Department forcing it to merge with the Missouri State Police.

CRITICAL LEARNING POINTS

1. Tell the whole truth the first time and every time!
2. Your goals for your department should be to become a high-trust and high-performance agency.
3. The best disinfectant is sunlight.
4. As humans, we will make errors, plain and simple. Don't compound the negative issue with lies or partial truths. Lying about the issue will only

make a bad situation a lot worse. Take responsibility for your actions and be accountable for all that you and your members do.

LEADERSHIP DISCUSSION QUESTIONS

1. Why is always telling the truth such an important personal behavior?
2. Describe and explain "the public's trust."
3. List one or two outward signs that the public you serve trusts your department.
4. Ferguson Police Department was used in the text as an example of an agency that lost the public's trust. Can you describe and discuss another example of a public safety agency losing the public's trust?
5. Has your department (or a neighboring one) been called into question about an issue of trust? If so, what was it? How was the issue resolved?
6. Why does the fire service need the trust of the public to perform its work?

PERSONAL LEADERSHIP PLAN FOR IMPROVEMENT

1. What behaviors do you demonstrate that are blocking you from telling the truth all the time?

____

____

____

____

2. What behaviors must you adopt to tell the truth, always?

____

____

____

____

3. List at least one training program you will attend in the next year to become a better leader and tell the truth, always.

4. List at least one leadership book you will read in the next year to become a better leader and tell the truth all the time.

5. List the names of the members of your department you will help mentor on telling the truth all the time.

6. List the name of one member you will ask to be your mentor on telling the truth.

7. Add one leadership preparation commitment you are willing to make in the next 12 months to "Tell the Truth All of the Time."

____

____

____

____

____

____

____

PRACTICAL APPLICATION AND RELATED CASE STUDIES

CASE STUDY 1

WANT TO FLAME OUT YOUR FIRE SERVICE CAREER? TELL LIES

YEARS OF BUILDING PUBLIC AND POLITICAL TRUST CAN BE SQUASHED BY SMALL COVER-UP FIBS, AS WILL YOUR CAREER.

Dennis L. Rubin

Tell the truth the first time. Tell the whole truth. Tell nothing but the truth. Tell the truth every time.

Never have wiser words been shared. What are the organizational rewards when we take this action? What are the pitfalls associated with lying?

As with all of the 13 career crushers, lying usually causes irreparable damage to your outfit and to a person's reputation. Making a mistake can generally be forgiven, if it is without malice and unintentional. However, lying doesn't go away without leaving a deep scar.

The fire-rescue service has the informal power to do the work at hand because it has the public's trust. We are vested with a wide range of sweeping formal powers that affect the lives of the residents and visitors to our communities.

The process of earning the public's trust takes place over a long period of time. Once earned, the community is usually willing to extend that trust until the agency breaks that bond through some errant action (which is usually stupid and highly publicized after the fact) that is real or perceived.

TRUTHFULNESS EQUALS TRUST

The focus of the leadership of any fire-rescue department should be to operate a high-trust and high-performance department that upholds the great tradition of service that we are known for delivering. Truthfulness is at the very core of the process that obtains and upholds the public trust. Effective leaders understand and fearlessly protect this earned, highly valued status.

We are a highly visible and highly scrutinized component of our communities. Our response equipment is brightly painted with reflective stripes; add sirens and air horns, and our every move is easily tracked by even the most casual observer.

Excluding HIPAA-protected information, all of the documents regarding the actions we take are available to the public. Most news outlets have reporters assigned to covering all aspects of public safety and they cover most of our activities.

Once a significant response occurs (measured by loss of life or large dollar loss), the incident will be gone over with a fine-toothed comb. Constantly remind your firefighters that we are being watched, recorded, viewed, reviewed, researched and tracked in just about every way imaginable.

Having the public's trust and support becomes mission critical. Lying will destroy that fragile trust quicker than any other factor. Lies will lead to mistrust and will crush community support of the department.

TRUST SCENARIOS

The best advice is to not engage in any dishonest behaviors. The inappropriate action always seems to be discovered at some point and the fibbing house of cards will collapse.

Consider this fictional scenario of a person who was injured in a serious automobile crash. With about 1.5 million such events occurring each year, it is easy to see that this is a realistic situation.

Because of the mechanism of this injury in this situation, it is necessary (following EMS protocol) to strip the patient's clothing to determine if there are unseen injuries. The person in need of your prehospital care is a teenager.

What will that teen's parents say when they find out? What will the community say about you removing this young person's clothing?

I submit to you that there will not be a discussion about that part of your treatment protocol, if you followed it correctly. To examine a patient for all injuries is part of the expected and accepted process of delivering evidence-based medicine.

ONE LIE TOO MANY

The lack of public comment and concern is only based on the fact that the member delivering the medical care and the department possess the public's trust. If that same situation plays out, and it is later learned that the EMT is a registered sex offender (there are about 750,000 registered sex offenders), there will be hell to pay by all involved—and, quite frankly, there should be.

We are empowered to do our job without question and with community support by virtue of holding the public's trust. Do not take this trust lightly or place it in jeopardy at any time for any reason. One single lie told by just about anyone who holds the public's trust can cause the system to come crashing down.

To put this career crusher into perspective, here's how it can negatively affect the individual member. If a firefighter has or would like to have police powers, associated with the position of fire inspector or fire investigator, be very careful about always telling the truth.

If a person holding these responsibilities fails to tell the truth (often described as lacking veracity, a more polite way of calling someone a liar), expect to be added to the Lewis List or Brady List. In essence, being listed means that you will not be allowed to testify in court or prepare public reports without disclosing that you had previously lacked veracity in some phase of police powers activity.

The focus on the witness stand will not be about solving the crime that was committed and determining guilt, but it will be about what the fire inspector or investigator did to be included on the Lewis or Brady list.

FINANCIAL BACKING

The last impact to consider is the overall, general community and political support for the organization. Most departments that have both the community and governing body's support are able to obtain the necessary resources to do their jobs.

Considering the costs involved in operating a department, all of the support the leadership can muster is needed to acquire and properly maintain fire and rescue stations and rolling stock.

If the public does not trust you to do your job, they will not trust you to spend tax dollars.

Always strive to be a high-trust, high-performance agency. The number-one factor in reaching and maintaining this lofty goal is...always telling the truth.

When a mistake is made, own it and be accountable for your actions. If it is an honest mistake, make efforts to correct it and offer a plan to improve the member and the organization.

Lying, on the other hand, will only make a bad situation worse. It seems like the best cover-up plans get foiled over time. So, tell the complete truth the first time, and tell it every time.

8

Do the Tough Stuff First

You will never reach your destination if you stop and throw stones at every dog that barks.

—*Sir Winston Churchill*

The choice of the day's work activities is typically made by the company officer at the beginning of each shift. In fact, the list of duties is generally left up to the station/company commander, with the exception of emergency responses and some planned events such as designated training exercises that are scheduled by a higher authority. Tough questions for the boss are "What does the daily activity plan look like?" and "What should be accomplished during the company's tour of duty for the next 12 or 24 hours?" These seem like such straightforward and simple questions, but the variance in accomplishments (lack of consistency) between shifts and officers is mind boggling.

Many companies will make the most of a shift or duty day, while others may not be as productive. Finding and maintaining a workload balance is a leadership art unto itself, as most supervisors quickly recognize. But one thing is for certain, if you tackle the difficult items at the beginning of the shift, they are exponentially more likely to be completed in a timely fashion. The time-tested adage, "Don't put off until tomorrow what you can do today," comes to mind.

There are many responsibilities and duties that are not pleasant or very desirable. As a firefighter, washing windows was one of those dreaded tasks I never looked forward to being assigned. In fact, I would attempt to swap house chores with another member (with little or no luck) whenever I was assigned to clean the windows. It was always much easier to complete the house duties that I enjoyed (like preparing the apparatus for response) and that seemed (to me) to make a functional difference. As a private (firefighter rank nowadays), I would even report for duty a bit earlier than my colleagues to ensure a better chance of being assigned more "compatible" house duties than those dreaded windows. However, I still occasionally found myself with an assignment that was simply no fun. The older members would make encouraging statements: "Take the bitter with the sweet, rookie." Of course, these were the same folks who would work their hardest to make sure all of the other, newer members were assigned the tasks they didn't enjoy completing.

If left to my own devices, I would have swapped duties or procrastinated and not started the hard stuff. Either the task would be assigned to the next shift or the windows would become filthy under the "Private Rubin House Duties Plan." Needless to say, passing off the work or not doing the required cleaning were not options, so I would reluctantly go about my duties when required to do the less glamorous tasks. You could only wait for a working fire or "megatrauma" code for so long before it was obvious you were slacking off on your assignments.

SOMETIMES THAT STUFF CAN RUB OFF

I always looked forward to getting to work early in the morning. If you reported to duty at just about the right time, you could avoid the horrible traffic jam and arrive early enough to be assigned the small line position (which meant you would be carrying the nozzle into the hazard zone with the company officer). Once in quarters and settled, the routine would be just about the same each day. I would place my turnout gear on the apparatus and return the member's turnout gear that I was relieving into the same boot locker I just left. Next, a quick check of the equipment that was assigned to the position that I was riding that day. SCBA check, booster tank water, hoselines properly packed and ready to go would all be on that list.

After the basics of assuming duty were complete, it would be time to focus on breakfast. For many years, I worked for a particular lieutenant who, once breakfast was over, would relocate from the dining/sitting/TV room to the company officer's desk in the back of the apparatus floor. Once seated, our lieutenant would read the newspaper from the front page to the obits. After catching up on the news, he would start working the crossword puzzle. Finally, he would cut the completed puzzle out to mail it in for the grand-prize drawing of $100 (a lot of cash back in the day). If you haven't guessed by now, the lieutenant's routine would take hours, leaving his four firefighters without supervision or direction.

I would love to tell you that we continued with housework or drilled on firefighting skills or perhaps pump operations, but that would be a fib. Just like the well-fed lieutenant, we all just wasted our duty time on playing cards until we had a run. I was very glad that we were a very busy company and calls headed our way regularly.

The point of this brief career flashback is to remind leaders that when you procrastinate and put the work off until later, it drains the enthusiasm, energy, and motivation from the members under your command. Remember, most folks look forward to and welcome mentoring and leadership. Keep this fact in mind as you plan out the work day. Don't forget to always include job readiness (training) in some form or fashion as part of the daily plan as well.

My lieutenant would make captain and be transferred out a few years later. His replacement was an amazing leader who had a great influence on my career. The new captain at Engine 10 was not a newspaper maniac or crossword puzzler. In fact, he was a firefighter's firefighter. We drilled every day for at least four hours morning and afternoon (Sundays and holidays were no exceptions). I learned more about my job from the new captain than anyone else during my formative years and became a journeyman firefighter. I was lucky that this captain was assigned to my beloved Engine 10 on Number 3 platoon.

BECOMING THE "DESIGNATED ADULT" ON SHIFT

As I became a little more senior in my position and started up that promotion ladder, the harsh reality of having too much work to complete in one shift became abundantly clear. The need to develop a priority system and put the shift into some type of logical order was apparent. For a short while, I would go after the low-hanging fruit (easy stuff) and completely fill my day. That seemed to keep the higher bosses as well as my crew happy. However, when mandated projects were slipping through the cracks and I was the one causing the less-than-desirable performance, it was pointed out to me. I took the heart-to-heart discussion in stride and committed myself to getting all of the work completed in the timeframe allotted. So, as a matter of personal survival, I started to do the tough stuff first, not just the fun tasks. What a valuable lesson that was: get the tough stuff behind you early in the shift and the rest of the day flies by. Life becomes a breeze.

As an example, later in my career as a chief officer, I hated dealing with the endpoint of the disciplinary process (employee termination). It always felt like the department had failed by hiring the wrong person in the first place or by not being able to reach the member in trouble with a substantial improvement program before it was too late to save that person's career. So my tendency was to leave this until the last possible minute. Oddly enough, waiting to deliver bad news (about discipline or anything) never made the information better or easier for me to deliver. It just extended the agony of being the bearer of the unpleasant facts of the process. It took a while, but I realized this was not helping anyone, and that handling the difficult personnel issues in a timely fashion (first thing at the start of the shift) was the only appropriate way to deal with this difficult duty.

THE BENEFITS OF THIS ADVICE

If you complete the tough jobs at the start of your shift, the day will get easier as you go. I understand that this rule is a blinding flash of the obvious, but it is important enough (and forgotten enough) that a reminder of its value and power is useful here. As a direct result of taking the tough issues head-on, you will develop a solid reputation as a person who is a high performer, dependable, and able to prioritize properly. Your reputation as a leader will skyrocket. Keeping the organization's needs ahead of your own by getting the difficult tasks out of the way early will speak volumes about your good character without you ever saying a single word.

If I had the fortitude to complete the less desirable or more difficult assignments before the action-packed events occurred, that made for a wonderful day at work. However, when I let things get out of balance (failed to handle the tough stuff early), difficult times were always just around the corner. I can't count the number of times I have completed budget documents on a weekend or stayed late on shift to complete personnel evaluation forms that were required for a member to receive a timely pay increase because an emergency event took up all of my shift time. Do the tough stuff first.

SUMMARY AND REVIEW

Set the shift's work priorities early in the day (or even during the previous shift). Make sure that all of the tough stuff is completed at the beginning of the work period. Be realistic and be flexible—the priorities of the day will quickly change with alert tones and response bells going off—but keep in mind that if you put off the important or difficult tasks, they will likely not get done.

I have had the wonderful opportunity to attend the National Fire Academy (NFA) on a few occasions. I am proud to point out that I am a 1993 graduate of the renowned Executive Fire Officer Program (EFOP). In the executive analysis leadership course, our cohort worked on a role-playing process that cast us in the roles of labor or management during a contentious contract pay reorganization process. Each side was given all the necessary details about the process and the ground rules that were to be followed by both sides of the table. The time element was that three afternoon sessions were planned at two hours each. Perhaps I don't need to mention the results, but we completed the work assignment with just a minute or two left on the clock, and that was with the federal mediator (the course instructors) prodding us to get the task completed. Reflecting back, we were not focused as a team or individuals on getting the tough stuff completed early in the process. The focus should have been for "labor and management" to engage in the heavy lifting (critical ties to be incorporated into the agreement as pointed out in the student activity) and not waste time. If our class would have tackled the tough stuff first we would have been able to complete the assignment within the time constraints.

Some of our duties are not pleasant ones, for instance, instilling discipline or completing employee evaluations. However, they are very important to the success of the organization. If you find yourself needing assistance to understand a process you are responsible for, reach out to your supervisor or a departmental expert in the specific area of challenge. However, you must supply the ambition and desire to do the tough work. You are the only person in charge of this part (enthusiasm) of the process.

The final thought is that if you are the leader and put off your duties, or worse yet, go to work to loaf and relax, the members will not be motivated. They will follow your lead and simply slack off and waste time. Stay engaged and keep the members productive, happy, and well-trained.

CRITICAL LEARNING POINTS

1. Plan activities out in advance to get the most out of your time at work.
2. The focus must be one getting the tough stuff accomplished first. There should always be accountability for completing all assigned tasks.
3. The leader's lack of motivation is contagious. Your laziness and failure to handle responsibilities will rub off on your staff. Be the leader—complete the work.
4. When all else fails, refer back to Rule 2: Lead from the Front.

LEADERSHIP DISCUSSION QUESTIONS

1. Describe a situation or two when you failed to get your assigned work completed. How did you feel about the experience? What steps or controls did you implement to resolve this issue?
2. Have you been assigned to a commanding officer that procrastinated? What was the experience like? Did you and your company rise above the issue and do good work without leadership?
3. What is the most difficult house chore or administrative responsibility that you must complete? Do you sometimes miss the assigned completion deadline for this assignment? What process do you use to keep on track with completing unpleasant assignments?
4. During my career experience, I describe the "new" captain at Engine 10 as being a mentor and effective leader. In your opinion, do firefighters look for those traits in an officer? Why? Why not?
5. Describe at least one situation where you followed poor leadership. What was the end result? What did you learn from this experience?

PERSONAL LEADERSHIP PLAN FOR IMPROVEMENT

1. What behaviors do you demonstrate that are blocking you from doing the tough stuff first?

____

____

____

____

2. What behaviors must you adopt to do the tough stuff first?

____

____

____

____

3. List at least one training program you will attend in the next year to become a better leader and do the tough stuff first.

____

____

4. List at least one leadership book you will read in the next year to become a better leader and do the tough stuff first.

____

____

5. List the names of the members of your department you will mentor on the skill of doing the tough stuff first.

____

____

____

____

6. List the name of one member you will ask to be your mentor on the skill of doing the tough stuff first.

7. Add one leadership preparation commitment you are willing to make in the next 12 months to become better at doing the tough stuff first.

9

Be the Customer Service Advocate

Opportunities to add value to what we do every day are endless at no additional cost to the department.

—Chief Bruce Varner, Santa Rosa Fire Department

No set of rules could be complete without discussing the need for and the delivery of customer service. It is amazing how times have changed, setting in motion a reexamination of how we treat and think about our customers. Some folks in the fire-rescue industry take offense to the term customer service, as though it is demeaning. One fire officer has stated that Wal-Mart has customers; the fire service helps victims. Therefore, along with our focus changing, we need to redefine our relationship with the people that actually pay us (if you're a career member) or fund the operation of the department (if you work for a volunteer organization). When I first started in the department, the emergency event was seen as the customer. It seemed painful and disrespectful to talk directly to our real customers (the people) who were having a really bad day.

In some ways, the people associated with the problem were seen as an impediment to resolving the emergency at hand. They were not viewed as customers who we needed to talk to or comfort during our operations. As a young firefighter, it was a lot simpler for me to deal with just the emergency event. I was taught that there wasn't a need to interact with the people who were in great distress. Perhaps the idea was that dealing with the harmed people was the job of the police. I would say that we went to great lengths to separate the people suffering through an event from the event itself. Looking back, I see that the customers who were in the greatest distress never received the attention they needed or deserved as our customer.

THE SMELL OF SMOKE

I can remember heading off to a car fire that occurred on a limited access highway one early evening. That was very exciting! The large, billowing column of thick, black, highly carbonaceous smoke could be seen for miles. As we drew close, the smoke's blackness consumed the little bit of daylight that was left. The traffic had been stopped by our police partners and all hands were focused on the action like a laser beam. We were riding in on the big red truck to save the day, just like it should be, textbook-style. Our 3 in. supply hoseline was laid to the fire hydrant access points on this highway. The 1.5 in. attack line was stretched just like we were taught. The operation was poetry in motion. At this fire, there was a big crowd of people (perhaps because of the time of day, the location, and billowing smoke cloud) and we needed to be perfect under such public scrutiny.

As the operation progressed, the fire was knocked down in just a few minutes, and the smoke's intensity died as the flames diminished. The layout person moved into position to open the hood with the assistance of the lieutenant using the flathead axe and Halligan tool. The tactics of the day were going well at this point. So far, so good, as the raging car fire was extinguished. After opening the hood, which exposed the bulk of the fire, the same two members moved to force open the trunk and passenger doors to check for fire extension in those compartments. Inside of four minutes or so, we had the upper hand on this once out-of-control blaze, and as always, it felt good to tame the beast.

The members of the engine company crew completed a thorough overhaul of this once nice-looking late model Chevrolet, making sure the fire was completely extinguished. Then, we (the members of the engine crew) cleaned up our tools and appliances, using the booster line, while we waited for the fire investigator to arrive. Our company officer spoke to the passengers of the barbecued automobile to obtain the basic information (name, address, insurance, etc.) for the all-important fire report preparation. Within the next 30 minutes or so, I was standing on the dirty 1.5 in. hose rolls on the back step (no jump seats in the early 1970s), headed back to quarters to get ready to respond to the next alarm.

The only thought we had about the people involved was to hope that they didn't get hit by other cars while they stood out on the high-speed roadway. The ambulance crew was always busy, and they didn't need any more calls that day. We only had a few seconds of discussion and interaction with the real victims. The last act of their fire department when we departed was to cover this stunned older couple in a cloud of diesel exhaust. That should be pleasing "customer service" to any taxpaying citizen! How could they not love the firefighters who had hardly a word to say to them and could not spare the time to help them out of a dangerous situation (standing on a busy highway)?

THE NEW WORLD ORDER

Nowadays, we have a completely different and more comprehensive approach to solving customer problems. First, we properly and quickly resolve any emergency situation by flawlessly executing the basics of our job (Rule 3). Once we have the upper hand at the situation (in this case, extinguishment), we must focus on the people who are indirectly and sometimes directly involved in and recovering from the emergency. This means the focus must become helping the humans who are impacted, and the only way we can do that is to open a line of professional and frank communication. Most folks do not have a lot of experience at calling fire-rescue services. This will likely be the first and, hopefully, the last time they are users of your community's response system. So, try to make a lasting great first impression with your skills, knowledge, abilities, professional demeanor, and of course, a great big scoop of customer service right on top.

Going back to the car fire case study, imagine if we had taken a few extra minutes after extinguishing the fire to pack up the couple and take them to a place where they could get the help they needed. The unlucky couple needed to be moved to a place where they would be safe from the oncoming traffic and could make a telephone call or two (this was before cell phones were in general use) to their insurance company, arrange to have their car towed and repaired, and finally, connect to a rental car to complete their trip and regain control of the situation. It would have been no big deal to help with any one or all of these items, and it is the right thing to do. It does not cost the department a dime to add this type of value with the services that we offer (fig. 9–1).

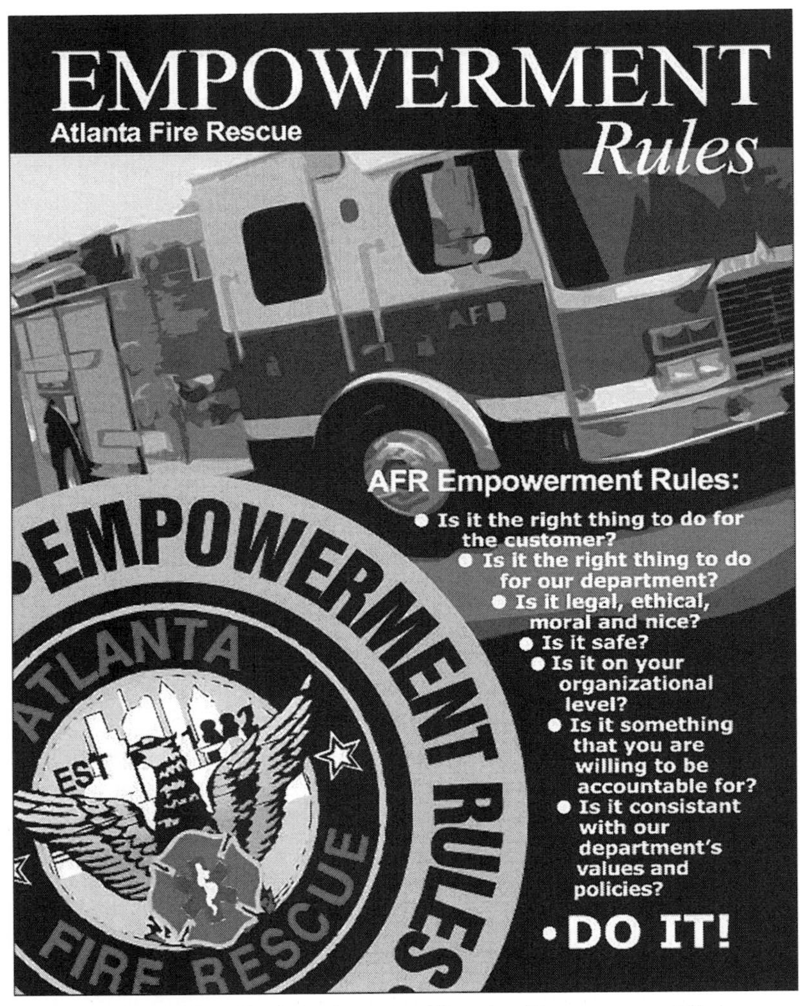

Fig. 9–1. Atlanta Fire Rescue Customer Service Empowerment Rules during my watch

DEER ROAST SERVED WITH A SMILE

One evening I was minding my own business, headed home after a full day at the office. It was such a busy place and there always seemed like there was something to do and deadlines to meet. As I was pulling into the grocery store to decide on the bill of fare of the evening, my departmental cell phone lit up. "Hello, Chief Rubin, may I help you?" The voice on the other end greeted me and asked if I could stop by Fire Station 16 for dinner. "Of course!" was my response. As it turned out, the dinner "inviter" was the truck company captain. He warned that the prepared meal was wild game and that dinner would be ready at about 6:30 p.m.

Armed with this information, I made my way to the firehouse, hoping that squirrel was not the planned dish. Although I had only been in town for a few months, I knew the way to this fire station, having been there a time or two. As I parked my car, I was met by the newest member of the station, who started washing my company car with great enthusiasm. Having the cruiser washed was never something I would have requested, but was always appreciated. It felt like the station was making sure I knew that the chief was welcomed, not feared. As the dust rolled off the black Crown Vic, I suspected there might be an issue aired at the dinner table. My hunch was spot-on.

As I entered, the captain walked up to greet me. I was a bit wary of what this new chief's "ask" might be. As I said hello to all of the shift members, I was wondering what could be pressing enough for me to get a last-minute invite to a fire station dinner.

Low and behold, a deer roast wrapped in bacon and coated with BBQ sauce and black pepper was being pulled out of the oven. It smelled and looked delicious. Thank goodness! I was relieved that I would not have to pretend to like squirrel meat. The meal was fantastic and incorporated perfect accompaniments. As I consumed the last few mouthfuls, I cleaned my plate with the warm homemade biscuits. Just when I thought the dinner could not get any better, a peach cobbler was placed on the stove top to cool off before serving. That's when it happened.

The captain spoke up before the dishes were cleared: "Chief, I have a bone to pick with you." *Here it comes*, I thought, and told him to please, fire away. After that great meal, I was ready for Station 16's questions.

The captain responded that he only had one question. I urged him to continue. "Well Chief, I thought that our department was going to focus on customer service issues, and that would be our new corporate value?" I assured the captain that his comment was accurate: the department's focus under my watch would be on customer service. "Well," he said, "Truck 16 had planned to install a lock on an elderly woman's home this evening. However, your deputy chief had sent the message down to not install the lady's lock."

"OK, please start this story from the beginning," I requested.

"Last shift we had a working fire in a Section 8 [low income and government subsidized] apartment complex. Among the job tasks Truck 16 was assigned to do was to check for fire extension from the unit that was on fire. We had to force a lock to gain access to fulfill our assigned duties. Thus, we destroyed the lady's door lock."

"Will the apartment management company replace the damaged deadbolt?" I asked.

"We reported it to the resident manager that night. The manager's response was we will get to it when we can, in a few days or weeks. I am uncomfortable that an elderly lady is living in an apartment without a lock on her door in a rough neighborhood." He completed his thought by explaining that the deputy chief found out their plan to install the lock and stopped the idea in its tracks.

Now I understood why I had been invited to the company's wild game dinner. I was needed to overrule the number-two boss of the department. Very early into my watch as fire chief, I had a tough decision to make. My response to the captain was, "Will you be taking the ladder truck?"

The captain replied, "Only if you let me." I asked if I could tiller the ladder and help install the lock. That made the entire crew of Truck 16 very happy, and a few minutes later, we were off to install the lock at the apartment. Once on location, the crew walked up to the second- floor unit and knocked. Once the tenant answered the door, the firefighters removed the electrical wire that had been twisted into some measure of security for the woman. As the real work of the installation proceeded, I had a chance to chat with our customer. She wanted me to know how much she appreciated the fire department and in particular, the captain, for replacing her lock.

What a neat night. I got to have a great dinner, tiller a ladder truck, help a customer, and have a great dessert. It sure beat cooking and cleaning up at home. I thanked everyone, expressing my appreciation for how the situation was handled. After shaking everyone's hands, it was off to the east side for a short night of rest and relaxation.

When I reported for duty the next morning, I barely got my key in the door before the deputy chief opened it for me. I could tell he was a cross between upset and worried. Not wanting to be ambushed at the front door of fire headquarters, I told him we could meet in my office after I handled a few critical administrative details, which included getting a cup of coffee. Once I was ready to discuss the issue, I called the chief into my office. As he walked in, he began explaining the decision he had made the day before. The deputy pointed out his plan was to get permission from city hall to install the lock at the apartment. I asked him how long it would have taken to come back from the city attorney's review.

Without pause, the deputy chief agreed that that may have been the wrong direction and stalled a great customer service idea. I asked him to get the receipt from Truck 16's captain and get petty cash to reimburse him for his personal out-of-pocket expenses. I then asked and the deputy agreed to head up a customer service committee that acknowledged and celebrated the fire department's customer service advocate for the month. To the deputy chief's credit, he did an amazing job at leading this committee, and we never looked back on me overriding his decision that day. Everyone was a winner at this end of this situation, because we took the time to provide excellent customer service.

SUMMARY AND REVIEW

How times change! It seems like life was so much simpler when we didn't have to interact so much with the people we serve. But when you think about the work that we do and the close connections we must build with the people experiencing an emergency, it just makes good sense to provide for the needs of our customers in a more direct and personal way. The above case studies simplify the general philosophy that needs to be applied. The opportunities to add value to what we do every day are endless.

I guess we have Chief Alan Brunacini to blame (actually, thank) for this dramatic change for the better. Once your organization truly embraces the customer service journey, the rewards will be amazing. They will first appear as kind notes and letters and ultimately as support in the all-important political arena. Never lose sight of the fact that the political arena is how we get our funding.

CRITICAL LEARNING POINTS

1. Be a customer service advocate whenever you can (fig. 9–1). This is the right thing to do, and it has many rewards for your agency.

2. In days of old, the actual emergency situation was viewed as the customer. However, a burning car does not pay you or agree to annual raises. Think about that.
3. Use the Momma Rule: What would your mother think about your treatment of others? and How would you want your mother treated?
4. Training, policies, and statements about customer service are important. However, walk the walk and talk the talk every chance you get as a leader (Rule 2).
5. Reward great customer service behaviors and celebrate them within and outside your organization.

LEADERSHIP DISCUSSION QUESTIONS

1. In your own words, provide an example of appropriate customer service delivered by your agency.
2. Why do you think it took so long for our industry to figure out the customer service component of business?
3. What is your opinion on the empowerment rules that were implemented in Atlanta? Is there a similar process in place in your organization? If not, should there be?
4. What guidance do you think is necessary to develop a list of customer service rules? What would your set of customer service guidelines be?
5. Describe and discuss a poor customer service situation that you were part of in your organization.
6. Describe and discuss an outstanding customer service situation that you were part of in your organization.

PERSONAL LEADERSHIP PLAN FOR IMPROVEMENT

1. What behaviors do you demonstrate that are blocking you from being a customer service advocate?

____

____

____

____

2. What behaviors must you adopt to be a customer service advocate in your department?

3. List at least one training program relating to customer service that you will attend in the next year to become a better leader for your department.

4. List at least one leadership textbook relating to customer service that you will read in the next year to become a better leader for your department.

5. List the names of the members of your department you will mentor on the skill of being a customer service advocate.

PRACTICAL APPLICATION AND RELATED CASE STUDIES

CASE STUDY 1

CASE STUDY: LESSONS FROM A FREAK-ACCIDENT RESCUE

A WELL-PLANNED AND EXECUTED RESCUE EFFORT PLAYED A LARGE ROLE IN THE VICTIM'S 'LUCKY' OUTCOME TO A BAD SITUATION

Dennis L. Rubin

Luck in the fire service has been best described as when opportunity meets training and preparation with the job being completed flawlessly. For this Tip of the Spear, we are going back to Monday, March 6, 2006. The location is Atlanta.

The weather that day was bright and sunny with very few clouds in the sky. The temperature would reach nearly 60 degrees Fahrenheit and the wind speed was just about 0 mph all day. What a picture-perfect, beautiful mid-winter day in the new South.

The shift started off quite tame. But at about 10 a.m. an alarm sounded for an unknown substance (white powder) in the Georgia State Capitol building.

The alarm would turn out to be a good-intent call with no service needed beyond perimeter control and sample testing the suspect white powder. Of course, the on-scene companies oversaw the evacuation of the Capitol to ensure that no one inhaled the unidentified suspicious white dust.

While I was checking in at the command post, the Georgia State Fire Marshal (Chief Alan Shuman) arrived. I updated him about the details of our progress to resolve this hazardous materials alarm.

The timing couldn't have worked out any better. As Chief Shuman gave the governor "all clear, under control" by phone, the two of us caught up and briefly talked about some recent arson-related issues in the city. A few minutes into the discussion, a single-engine alarm was dispatched that caught our interest.

THE CALL

Engine 29 was dispatched to a house where the front porch had collapsed, trapping someone. It was right at 11 a.m. Without any details about the collapse, Chief Shuman and I went back to discussing the arson issues. As Engine 29's company officer called in his brief initial report, it was obvious that this was going to be a complicated rescue.

The initial incident commander described that a concrete front porch had collapsed into the basement of a home with a man trapped under a large amount of debris. He went on to say the man's wife had entered the collapse zone to assist her husband and was in imminent danger. The next radio transmission was a call for a significant amount of help in the likes of a Georgia Search & Rescue Task Force (GSAR).

The man trapped was making the final preparations to take his wife on a family vacation. As he walked out his front door to load their vehicle for travel, the approximately 10×12 ft. front porch collapsed into a void space below.

The empty hole was left under the porch area during construction at the same depth as the finished basement and footers. That's right: all of the supporting dirt had been removed from under this front entranceway. The only structural components that supported the concrete slab that made up the porch stoop were ½ in. steel re-enforcement rods. Over many years, the steel rusted and corroded, allowing the porch assembly to go crashing about 20 ft. into the void space underneath.

RESCUE PLAN

It took everyone a minute or two to fully understand the level of danger that this situation posed, as well as how this accident occurred.

The homeowner was no longer headed on vacation, but was trapped under a large amount of debris with multiple systemic traumatic injuries. Once his wife realized what happened to her husband, she called for Atlanta Fire-Rescue's help and entered the deep hole to assist the unconscious and unresponsive man.

When the size-up was completed and the incident action plan developed, the first step was to remove the unharmed female. The technical rescue members were able to remove her over a ladder that was placed into the collapse zone without entering the collapse area.

The next step was to stabilize the dangling concrete slabs and soil that surrounded the victim. While scene stabilization got underway, the engine and truck company members performed the necessary support functions for a cave-in response, such as ventilation and air monitoring the opening, to determine if it would be safe for the rescuers to make entry.

The fire-rescue responders ensured that the building's water was shut off. This was to prevent water from leaking into this already compromised collapse zone, causing a secondary collapse. Next, ground ladders were placed in the opening to help the technical rescue members again access and remove the patient from the pit.

The ground ladders helped stabilize this area for the victim. They served as a makeshift trench box to minimize further harm to the trapped man as the team secured the caved-in area.

To be clear, this was not an active construction site. There was no digging or heavy construction equipment of any kind when the cave-in occurred. Stabilization consisted mostly of securing the dangling overhead concrete slabs and the earthen walls of the pit with pneumatic shores and preconstructed strong-backed plywood sheeting.

The GSAR team entered the hazard area and immediately provided advanced life support protocols to the injured man to control the profuse bleeding and to maintain a patent airway.

Many elements of the emergency-response system were responsible for saving a human life that day, including the communications centers, police who provide scene control, a private EMS service that dispatched an advanced life support ambulance, an aero-ambulance (helicopter) that was summoned, and finally the critical-care givers at the hospital.

TIPS OF THE SPEAR

Several members distinguished themselves above and beyond the call of duty at this alarm, earning their Tip of the Spear acknowledgment. Capt. Steven Woodworth and Lt. Jason Whitby were the rescuers who located and removed the seriously injured man. These two dedicated fire-rescue officers were an intricate part of GSAR 8 of Atlanta. In fact, both were working extra that day, conducting GSAR training when this call was dispatched.

Once the victim was packaged and removed for transport, the next step was to provide critical advanced life support care. Once removed from the hazard zone, the patient's bleeding increased remarkably. The next Tip of the Spear goes to Lt. Jimmy Gittens, who on this day was acting as the city-wide paramedic supervisor.

Lt. Gittens helped the on-scene medics to start two IV lines to replace the tremendous amount of blood being lost. Further, there was a great degree of difficulty intubating this critical patient. In just a few seconds, Lt. Gittens had the breathing tube properly placed and delivering oxygen. Finally, he took firm control and reminded all of us that this was a "load and go" patient. In the next few seconds, our patient was placed in a waiting ambulance helicopter.

Of notable interest was that a mature (35 ft. or so) conifer tree at the end of this cul-de-sac had to be removed to establish a safe landing zone for the aero-ambulance. The remaining Atlanta firefighters made quick work out of falling this large tree. In fact, once on the ground, they cut the tree into manageable fireplace-sized logs and neatly stacked the fresh cut wood and limbs out of the way of vehicle traffic. The members swept the area clean of all sawdust and debris.

Because of these firefighters' effort, the helicopter landed and took off without problem. I expected to get complaints about removing such a beautiful, healthy tree, but I never heard a negative word from this community about it.

EXCEPTIONAL COMMUNITY SERVICE

The final Tip of the Spear goes to Capt. Byron Kennedy, who was assigned to the Office of the Fire Chief as the public information and community relations officer. Very early into this event, it was obvious that the victim's wife was psychologically traumatized by what had happened.

Capt. Kennedy took the responsibility to be the customer liaison and advocate for this young lady, helping her in any way that he could. As the helicopter headed off, Kennedy ensured that we cleaned up the interior of the home.

Most of the operations were conducted inside the victim's home as well as providing the final ALS ground medical treatment. The blood and bandages cleanup was a daunting task, but had to be tackled before significant damage occurred to this home. Next, Capt. Kennedy helped the lady secure her home as best she could, then drove her to the hospital. Capt. Kennedy stayed with the lady until her husband finished surgery and was placed into a hospital room.

> Over the next few weeks, Capt. Kennedy took the lady back and forth between home and the hospital as she requested, as well as helped with any family issue that he could.

Now, let me explain why I say this family was lucky. Luck in the fire-rescue business has been defined as when opportunity meets training and preparation, and the job is completed flawlessly. Using this simple but accurate definition, this was a lucky day for the patient. Every aspect of the heavy and tactical rescue system was deployed with tremendously effective results. Perhaps the best-trained Atlanta rescue technicians responded to this call for help. Watching Lieutenant Gittens work his paramedic magic in such a calm and professional manner was a career highlight for me. As the final steps were being made for transport, I did not think that there would be a positive outcome. But because of the outstanding prehospital care, I can assure everyone that a life was spared on this day.

Watching Captain Kennedy help the victim's wife regain her composure and focus on how to effectively support her partner was a lifetime treasure. The follow-up support and compassion that he displayed made me proud to be a small part of an outstanding organization, where the tradition of great service continues today. Every member who attended that alarm did their job and a little more to make sure that the outcome was a positive one. It is true: training and preparation met opportunity head-on at Roxboro Point that day. I could never thank all of the responders enough for this and many other years of great performance.

In Fall 2006, the department was pleased to host this family as our honored guests at the annual awards ceremony. The ceremony started with a video clip that documented this successful rescue to the eventual recovery and rehabilitation of the once critically injured man. The video clip was amazing; however, it paled in comparison to expressions of gratitude from this very appreciative, loving, and thriving couple. As part of the department's acknowledgment and thanks to each member who worked this incident, a dress uniform ribbon was presented along with a wall certificate describing the outstanding performance.

What a wonderful day at work when everything goes right, and there is no doubt that a life was saved and a family made whole by the fire department's actions. Until next time, be safe out there!

CASE STUDY 2

PREVENTING HARM, ONE HOME AT A TIME

Dennis L. Rubin

After a long absence (about 28 years' worth), it has been great to return to Washington, DC and once again put on a DC firefighter's uniform. After I completed six years of service at Engine 10 in the 1970s, I had a wonderful opportunity to get out and see the "fire rescue service world," working for several different communities. That's right, I am the guy who's teased about not being able to keep a job; what a great set of experiences for me and my family.

In my absence, so much in DC has changed and so much has stayed the same. When I left our nation's capital, we had the Washington Senators Major League Baseball club; now it is the Nationals. The fire department operated with a three-platoon system; now it has four shifts. Engine 10 was the busiest in the city, and it still is; now even more so. We saw a considerable number of fire deaths, and, unfortunately, we still suffer with losing 10 to 15 civilians every year to fire. There has to be a way to change this grim statistic and lower or eliminate our citizens and visitors from needlessly dying in their homes. This column will discuss an initiative that was in July 2007 in the Deanwood section of our city. The program is titled SAVU, which is pronounced Save You and stands for Smoke Alarm Verification and Utilization, and will go a long way to curbing the fire deaths that we respond to each year. Perhaps this program can be implemented in your community and reduce harm, human suffering, and loss of life.

TRAGEDY STRIKES

Just a few days after returning to work at DCFD, we responded to a house fire on Minnesota Avenue, Southeast. The call came into our Office of Unified Communications at about 2:00 A.M., but the caller provided the wrong address. In fact, several more folks called in this raging blaze, giving the incorrect address several more times. As luck would have it, DCFD Engine 15's route of travel to the dispatched address went directly in front of the rowhouse that was burning. Thank goodness the response time was just under six minutes that night because the customers in that house would need all of the help that we could provide to avoid a very tragic outcome.

Upon arrival, Engine 15 provided a brief initial report that included the fact that people were trapped and requested a working fire assignment (which adds a sixth four-member engine, a third five-member ladder, another two-person command team and support units such as the air unit, fire investigations and the shift commander) to the alarm. As Engine 15 made the staircase with a ½ in. handline to attack the fire and begin a primary search, one of the truck companies stood a portable ladder to help some of the family members down from a second-floor bedroom.

Several folks were assisted to the safety of their front yard and then into a waiting ambulance. As emergency medical care was being provided that night, it was learned that the five-year-old daughter was not so fortunate and remained in her bedroom. The interior attack group, led by Engine 15, discovered the little girl on the second floor, severely burned and lifeless, just minutes into this firefight. Tragically, the little girl would not survive this fire that consumed the top floor of her home.

It is difficult for firefighters to deal with any civilian fire deaths; however, it is even more disheartening to realize that the person who didn't make it was an innocent child. That being the case and an overwhelming need to do something about this situation, the concept of SAVU was developed and presented to the devastated neighborhood at a Sunday-afternoon press conference that included Mayor Adrian M. Fenty and Council Chair Vincent Gray. The press conference was held on the front steps of the burned-out home where the precious little girl lost her life. It was well attended by the community and the media.

SAVU PROGRAM IS BORN

The plan was simple and potentially very effective, but it would take a lot of hard work to implement. The written goal of the plan is simply to prevent harm in our community. Knowing that about half of the homes in our city don't have a properly installed and maintained smoke alarm, DCFD has agreed to take on this huge challenge. The functional elements of the program are to change the direction of the community smoke detector efforts from a smoke detector giveaway to knocking on doors and installing and repairing existing devices, a remarkable and significant change of direction.

The next several months were spent resolving the many details (as Assistant Fire Chief of Services Tom Herlihy always points out, "the devil is in the details," and that was a fact for this program). The General Council had to get approved a standard release form to let our members to go into homes and install the devices. Next, we resolved the operational components from developing installation kits to providing personnel to do the work. The information technology folks helped by taking a section of the city and dividing it into six segments in residential neighborhoods. Each section would have about 100 addresses to be visited on our Smoke Alarm Blitz day. The idea was that an operations company would be assigned to each of these areas. The company would be split into two units. Four community volunteers would be added to knock on the 100 or more doors on blitz day. Assistant Fire Chief of Operations Larry Schultz made sure that came together flawlessly.

Next, we had to fund this mission-critical event, which was not in our budget. This would be the part that would take some creativity and hard work. Several old friends once again came to the rescue. The necessary money was donated by a very supportive business and several other great companies helped us with the smoke/carbon monoxide alarms. As the clock and calendar seemed to work against us, the required resources came together just in time to implement our SAVU program with only a few days to spare. It seemed like only the spray test smoke was going to elude us on show day, but lo and behold, with the help of overnight drop shipment, the cans were present and accounted for to test the newly installed devices.

Perhaps the most impressive component of the planning was the development of a Type 4 incident action plan (IAP). This plan was six pages long and included every detail, from the operational period goals and objectives to safety to logistical support to assigned radio frequencies. I must commend Battalion Fire Chief William Flint and his companies that day for far exceeding my expectations. July 21, 2007, will be remembered for the kick-off of this successful program.

AN OLD FRIEND RETURNS

As we were planning to start this major initiative, I happened to see a longtime friend and former member of our department, Dr. Burton Clark, at his office at the National Fire Academy, where he is chair of the Executive Fire Officer Program. I remembered that Dr. Clark (then, as he would remind me, Private Clark of Engine Company 24) was responsible for developing our first smoke alarm program. Burt was then and is now on the leading edge of our industry and always out in front. So, remembering that he got DCFD originally started on this journey, I asked him to assist us by saying a few words about the agency's history and commitment to preventing harm in the District of Columbia. Without hesitation, Burt agreed to help us without condition. I asked if he needed to check his very busy schedule, and he assured me that he would clear it if necessary. No doubt about it, Dr. Clark was and is committed to this program and to our department.

As all good leaders do, Burt asked what my expectations of him would be that day and how he could help. I was hoping that he would make the trip to Northeast Washington that day and reflect on our past. I was sure that if the members and our elected officials were to hear how we got started so long ago, there would be even a stronger commitment to success.

Dr. Clark once again raised the bar by pointing out that just saying a few words [was] not enough, he was staying that day to help us by installing the devices. "Private" Clark was assigned to work with Engine 17 ("The Protectors of the Holy Land") for that event.

SHOW DAY

When Burt agreed to be a partner, helping to kick off the SAVU program and with the financial support issues behind us, I started to breathe a little easier. I now believed that SAVU was going to happen and be successful. The morning was full of activity as the final parts of the plan were put into motion. I arrived at the appointed location at 7 o'clock only to find that most of the heavy lifting was completed. Our command post vehicle was the center of attention, with many folks assembled around to obtain supplies and orders for the day. In front of the command post were a podium and stage for a 10 a.m. press conference to kick off the event. The smell of pork and chicken barbeque filled the air. I was grateful to learn later that the IAP included a world class lunch prepared by DCFD's barbecue team (that is correct; we have a group of firefighters who tour this nation and compete for various honors for their cookery).

Several other elements were taking place while I marveled at the efficiency and effectiveness of the command team and various support players. Among the activities were sign-in and supporting paperwork to let the volunteers help and ride out with their prospective companies. Next, the unit leaders (officers and drivers) were in training and operations briefings. For about 30 minutes, company officers and drivers were updated on the details of the SAVU program and given their street assignments. It was soon after that the volunteers were paired with their apparatus for the final unit-level briefing.

At the stroke of 10 a.m., the press conference started. In all there was our mayor, council chair and the ward's council member in attendance. They spoke to all of Washington's media agencies about just how critical this program is and that everyone must have a properly installed smoke alarm. Another unexpected highlight was the fact that all three elected officials agreed to provide $300,000 of support to operate the SAVU program for the next year. Among the improvements that we hope to reach is adding a part-time position to manage this high-profile, community-friendly program.

Along with our governing body, we had several community leaders to include the parents of the five-year-old girl who lost her life. The girl's mother gave a powerful speech (fig. 9–2). The final speaker of the day was Dr. Clark. He gave a resounding speech that sounded more like a testament to the department for being committed to this community outreach. The history lesson about the smoke alarm promotional program for the 1970s and a bit of insight about even the department's reluctance to trust the devices 30-plus years ago was thought-provoking. Then, to have Burt lead by example and climb aboard Engine 17 to work as a smoke-alarm installer for the day was nothing short of an exclamation mark for this program kickoff.

Fig. 9–2. Washington, DC, Fire Chief Dennis Rubin left, and Mayor Adrian Fenty stand behind the mother of little A'sia Sutton, who died in a house fire on Minnesota Avenue, Southwest, as she discusses the importance of having a working smoke alarm in homes at DCFD's Smoke Alarm Blitz. Photo by PIO Alan Etter, DCFD.

A PEEK INTO THE FUTURE

As the 3 p.m. approached, we had reached every one of our objectives. More than 200 alarms were installed (the units were either smoke and CO devices or the newest—a talking detector that provides detailed instructions using the child's Mom or Dad's voice). Over 600 homes had to be reached. Although most of the customers were not at home, we left a calling card that explained the program and a number to call to have a device installed or help to repair an existing one. The number-one goal was to operate in such a way that no member or volunteer would be hurt, so another goal reached with no injuries.

Next, we wanted to make sure that the entire event was well documented to include media coverage at the opening of the event and to record all of the residences that were checked; with a lot of help from our IT staff, this was accomplished. Finally, all workers were to be fed and thanked for the extra effort, and that was handled by Assistant Fire Chief of Planning & Preparedness Brian Lee.

The decision was made at the end of the day (and supported by our labor association, IAFF Local 36, fire and EMS administration, community leaders and the elected officials) a Community Smoke Alarm Blitz would be held every third Saturday of the month in a neighborhood that experienced significant fire-loss history. The second blitz is being planned, and other developments are in process. The hope is that 10 more companies will go out and install 10 units each in their first-due areas. When we add this component to the SAVU program on smoke-alarm blitz day, we will install another 100 detectors. Finally, we have added a web link and a hotline telephone number for families to call to get an alarm installed on demand for families that are not available on Saturdays. We will be watching the civilian fire-death statistics closely to determine the effectiveness of this program, making the needed changes to make it successful. I am sure that it will make a difference in preventing harm in our community.

10

Demonstrate a Passion for Health and Safety

We as firefighters can't help anyone else if we need help ourselves.

—Chief Mark R. Nugent

Great leaders must ensure that all firefighters go home at the end of an alarm or shift. This rule is the most important one in the entire list of leadership rules. The primary responsibility of all officers is the safety of the members under their command. Our members must return to their families and homes just as they arrived at the firehouse, before the response to an event or the completion of a shift.

The fire-rescue department exists to help others during their times of need. The notion is that we must be able to help folks when, where, and how their situation demands. We perform on the streets of our communities as the emergency dictates. For the fire department, one main solution tends to be fire extinguishment, because when the fire is extinguished, the entire scenario will get better. Sometimes, however, holding a defensive position until the police can control a hostile person may be the better option. Always use a risk-versus-benefit model to decide the proper level of engagement and firefighter risk. If your firefighters are not physically and mentally at 100% (due to injuries sustained during the shift, for instance), you will fail at the mission you are sworn to carry out: protecting the public we serve and preventing harm in our communities. Without the required human resources to perform that work at hand, you will not be successful. Our industry is and will likely always be highly dependent on people (trained and qualified firefighters) to carry out the required tasks to save lives, protect property, and help secure our communities (figs. 10–1 and 10–2).

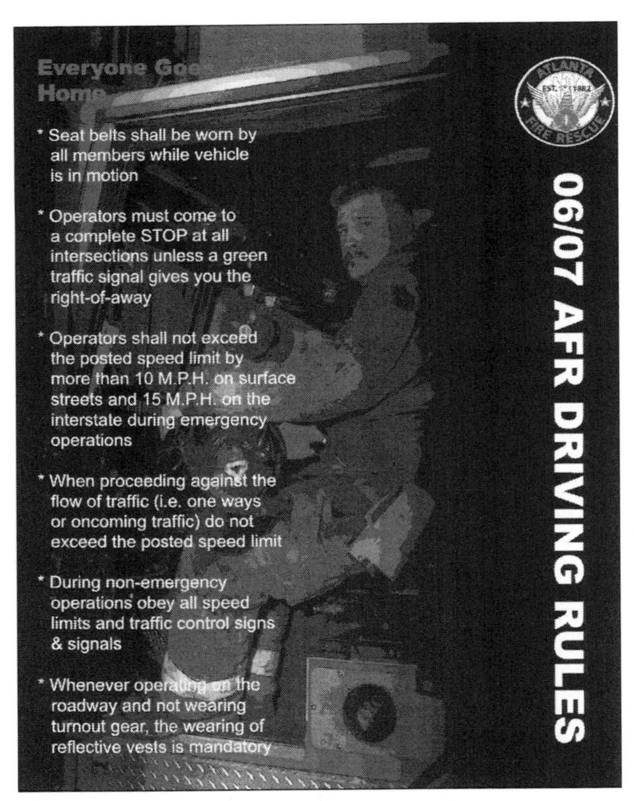

Fig. 10–1. Atlanta Fire-Rescue driving rules part one (rules to live by)

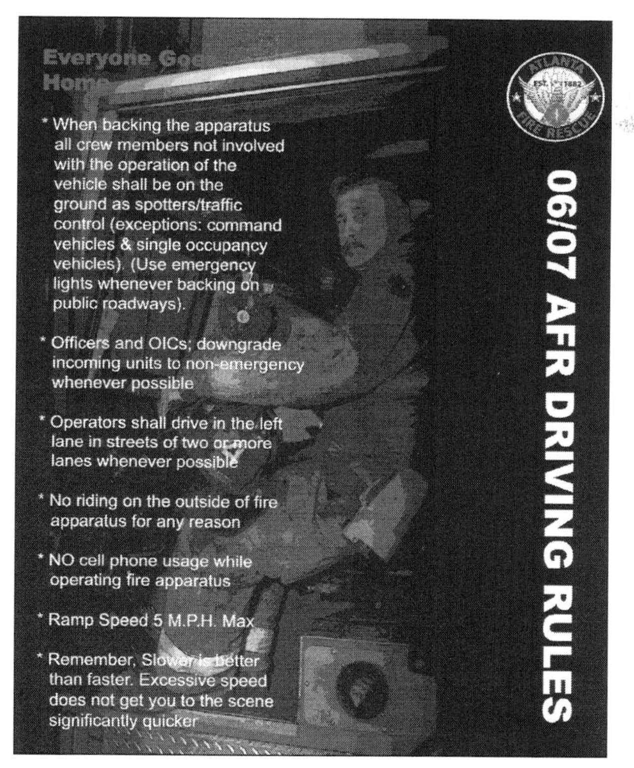

Fig. 10–2. Atlanta Fire-Rescue driving rules, continued

CAUGHT UNDER THE WHEELS: THE SANDY LEE STORY

One of the most applicable case studies I can share is the tragic event that occurred on September 28, 1982. A young, capable firefighter was on duty assigned to a ladder truck in Prince George's County (MD). During the evening of the shift, a box-alarm assignment was dispatched for an apartment house on fire. Firefighter Sandy Lee quickly reported to her assigned position on the right jump seat of Truck 22 in preparation to respond to this call for help. However, Lee skipped the step of donning her turnout gear and failed to sit down in the jump seat and use the seatbelt that was provided for her personal protection.

As the response started and the ladder truck began to move, Lee's turnout boot fell from the jump seat onto the concrete ramp in front of the fire station. (This was back in the days of three-quarter boots and open jump seat areas.) A natural reflex was to reach for the boot as it tumbled away from her. In a split second, Lee fell from the vehicle onto the apparatus apron. With the siren and air horns sounding, her screams for help and her request to stop the vehicle went unheard. As you might have guessed, she was struck by the tandem rear wheels of the ladder truck and dragged more than 30 feet across the front ramp, sustaining severe damage to her body. This event would change her life and career forever.

Lee was first transported to the Prince George's General Hospital (PGGH) Level 1 Trauma Unit for stabilization. Soon after being stabilized at PGGH, Lee was flown to Baltimore's Shock Trauma Unit. The staff there was able to somehow miraculously save her life from such devastating, crushing injuries. Lee would require dozens of operations and hundreds of pints of whole blood transfusions to recover. It was a miracle that she did somehow survive this multisystem traumatic ordeal.

As her physical and mental healing process started, Lee developed an amazing training program she called "Caught Under the Wheels." Retelling her story scores of times, Lee made firefighters think about the importance of following the department rules (donning turnouts before the response, as well as being seated and belted before the fire apparatus leaves the station). She would emphasize to always use seatbelts before the vehicle starts into motion. There are several very powerful training videotape presentations that Lee appears in, with the goal of making sure this information is not lost on future generations of firefighters. (This videotape training program should be mandatory for all firefighters to view. Call me old school, but I know that hard-learned lessons are sometimes foolishly forgotten over time without refreshers.)

Lee takes full responsibility for her actions. She informs listeners that because of her accident, the ladder truck she was assigned to that night never made it to the apartment fire. Further, the medic unit and the battalion chief assigned to assist at the apartment fire also never made it there: they were all assisting with the injured firefighter who was down on the front ramp of Station 22 that night.

Lee will always hold a special place in my heart and thoughts. This is not just because she sustained career-ending injuries and endured immeasurable pain and suffering that night while protecting her community. In addition to being such a courageous leader and hometown hero, Sandy Lee developed a behavior-modifying training program, and she did something few folks do today at any level of our society—she took personal responsibility and accountability for her actions.

I think of Sandy Lee often, and I shall never forget the sacrifices and contributions she has made to the American fire-rescue service. When I discuss firefighter safety, I use this example to help folks think through the fact that you can't help anyone if you are hurt. Hopefully, you have never had a member injured during an operation. The folks who have had this unfortunate event occur can fully appreciate my next thought. The entire company you are working with goes out of service when a member is injured. In fact, if the injury is moderate or significant, other companies will be called in to assist by providing medical aid and transportation to the downed firefighter. When a firefighter is injured, this can actually be more difficult to resolve than the situation you were called on to handle in the first place. Because of many factors (mental distraction, reduced on-scene personnel, and fewer available apparatuses), the initial alarm becomes more dangerous for everyone still operating at it, not to mention our customers and their property.

READ-LEARN-STUDY

There are dozens of great firefighter safety and survival programs readily available. The National Fire Academy (NFA) is a great place to start or continue your lifelong educational commitment to firefighter safety. You should consider taking the NFA's on- or off-campus firefighter safety programs. Each day, more information appears on the web and is increasingly easier to access to improve our operations from a member-safety standpoint. There are great textbooks available that discuss just the topic of firefighter safety. I would urge you to be a lifelong learner and reader of all types of fire service materials and information. (*Rube's Rules for Survival* (2013) is one of my favorites.) The need to learn about how to protect yourself and your partners cannot be overstressed. All actions that can be taken to avoid the high likelihood (about 100,000 firefighters injured per year) of personal injury should be taken.

RISK MANAGEMENT MATTERS

Whenever members are placed in a hazard zone of any type, the major focus must be on risk management. Risk management is a very interesting topic. It is a simple process and is easy to understand. It is relativity easy to read and write about the RM process. However, applying all aspects of the concept proves to be a robust challenge. The concepts are well documented, with credit due to Chief Alan Brunacini for his work and research on this topic. I have learned from Chief Brunacini that there are three major factors to consider when implementing the risk-management model at any and all hazardous operations.

First, we should respond to the hazard with the mindset that we are willing to engage in a high level of risk if there is likely to be a high level of reward for our efforts. For example, a primary search-and-rescue (SAR) operation is worth the risk to members if there is likely to be salvageable occupants (people who are alive and will still be alive when we reach them) in areas we can safely reach. Next, we should be much more measured with our level of risk taking if the benefit has been minimized (for instance, at a fully involved fire where there are likely no salvageable occupants or property). Finally, when the situation points to little or no benefit to be gained by our actions, do not take a risk (for example, do not conduct primary SAR operations at confirmed vacant properties).

The best way to remember and describe this process is with the following three often-cited statements (often attributed to Brunacini):

- We will risk a lot to save a lot.
- We will risk a little to save a little.
- We will risk nothing to save nothing.

Now, let's review the factors that make the risk-management system difficult to use effectively. Once the correct control measure is selected, we have to remember that this process will always choose life safety (firefighter and civilian lives) over property conservation, which is sometimes a difficult concept for firefighters, who are trained to put out fires—all fires—and protect property.

Next, the risk-management model is implemented in a highly calculated manner. That means we must use all of our training, education, experience, and operational policies to work safely at any event. If the hazardous zone we are entering is a house on fire, here are a few of the policy, size-up, or training items we must consider in our risk assessment:

- Physics and chemistry of combustion
- Correct use of personal protective equipment (PPE) and self-contained breathing apparatus (SCBA)
- Appropriate ventilation and control of the flow path
- Hoseline placement and movement and water supply
- Need for and potential effects of forced entry
- Need for SAR operation;
- Placement of ladders
- Incident priorities

- Correct and efficient use of tools, such as the thermal imaging camera (TIC)
- Fire extension
- Radio operations
- Location and efficiency of command and control

These and many, many more functions need to be factored in to the risk assessment. I am sure you get the idea. There are a lot of moving parts that must be considered to properly attack a structural fire when we are inside the hazard zone.

Finally, as if this abbreviated list of things you must take into consideration is not enough, the last component of risk management is properly controlling the event. This is referring to properly implementing an incident command system (Blue Card, National Incident Management System [NIMS], Fire Ground Command [FGC], etc.).

HAZARD ZONE ACCOUNTABILITY

The last safety-related item that will be covered in this text is hazard zone accountability. Accountability is the process of keeping track of everyone at an incident to ensure their safety and survival. The most appropriate way to do this is to always be able to answer these five simple questions:

1. **Who are the members in the hazard zone?** Knowing them by name is preferable, but at least know the number of members and the companies to which they are signed. For example, Engine 10 has five members and Truck 13 has six members on duty and operating at the event.
2. **Where are your members operating?** Three members of Engine 10 have been assigned to Division 2 (on the second floor), one member is at the pump panel, and the fifth member is at the hydrant. Truck 13 has divided up into two three-person SAR teams above the fire on the third floor.
3. **What actions (assignments) are the members engaged in?** For instance, you know that you assigned Division 2 to conducting the fire attack on the second floor.
4. **What conditions are the members working under in the hazard zone?** Is there enough air to make a round trip (escape when it is time to exit) and return to fresh air?

5. **Are there clear exit pathways?** Keeping at least two separate exit paths clear and functional is a very important task for both the IC and the safety officer. Sometimes only one path of exit is available. If so, that pathway must be protected and kept usable until the operation is concluded.

SUMMARY AND REVIEW

Please make a concentrated effort to learn more, all of the time, about this critical topic. Be a lifelong learner and continue to add to your body of knowledge relating to firefighter safety and survival. Work toward and maintain the various levels of scene safety officer certification when time and resources allow. The entire operational process depends on you being capable of performing your assigned duties. All officers should obtain and maintain the national incident safety officer certifications to perform their duties correctly. You owe it to your family, yourself, and your department to come home after every run or shift. If you are a company or chief officer, your responsibilities in this process (member safety) exponentially increase. Never forget your first duty as a leader and protector of your people.

Please be safe out there!

CRITICAL LEARNING POINTS

1. The safety of members is key to the success of all incidents. Firefighter Lee said it best: "You can't help anyone if you are injured yourself." Injured firefighters at hazardous events drain resources, making all other members less safe.
2. Be a lifelong learner.
3. Risk management must be understood and implemented at every hazardous event.
4. Risk a lot to save a lot, risk a little to save a little, risk nothing to save nothing. Always choose life over property and take risks in a highly calculated and controlled manner.
5. Know, understand, and implement the five guidelines of hazard zone accountability:
    a. Who is operating?
    b. Where are they operating?
    c. What are they doing?

d. What are the conditions?
e. Are there two or more methods of egress?

LEADERSHIP DISCUSSION QUESTIONS

1. List and describe four elements that affect firefighter safety and survival.
2. Who has the ultimate responsibility for your safety?
3. What are the three major tenets of the risk-management model? What are two controlling factors that must be considered whenever you select the proper risk management model?
4. Why is it necessary to have at least two paths of exit from a hazard zone? What are the methods that can be deployed to protect pathways of exit?
5. List and discuss each of the five elements that make up the personnel accountability report (PAR).

PERSONAL LEADERSHIP PLAN FOR IMPROVEMENT

1. What behaviors do you demonstrate that are blocking you from demonstrating a passion for firefighter safety?

____

____

____

____

____

2. What behaviors must you adopt to demonstrate a passion for firefighter safety?

____

____

____

____

____

3. List at least one training program relating to this topic that you will attend in the next year to become a better leader by demonstrating passion for firefighter safety.

4. List at least one leadership textbook relating to this topic that you will read in the next year to become a better leader by demonstrating passion for firefighter safety.

5. List the names of the members of your department you will mentor on firefighter safety.

PRACTICAL APPLICATION AND RELATED CASE STUDIES

CASE STUDY 1

CREW RESOURCE MANAGEMENT–PART 6: CRITICAL DECISION MAKING

Dennis L. Rubin

After more than 30 years of utilization, the incident command system (ICS) has become a way of life for most North American fire and rescue agencies. Some departments and members are better at applying the command-and-control process than others, but all in all ICS has become an industry standard.

NIOSH, NFPA, and all other regulatory groups are very clear in their message: the use of the ICS is not optional. When this process first evolved from the California FIRESCOPE project and Phoenix Fire Ground Command program, ICS had the potential of being a pivotal/historical change in how we do business.

ICS could have ended up as a flash in the pan or hitting the "round file" in due time, but history reflects that our business took to ICS like ducks to water. It would be unimaginable to turn our backs on a process that has tremendously enhanced firefighter safety while allowing fire departments to be more effective in solving our customers' emergencies. What a huge, positive impact that the ICS/ IMS has had on our members and customers when used properly and consistently.

On behalf of our industry, thanks to Phoenix Fire Chief Alan Brunacini and the FIRESCOPE cities for having the vision and conviction to make such a sweeping, positive operational impact. With celebration and acknowledgment, it is now time to look to the future and be a part of molding the next organizational change. I am convinced that the next logical step in the firefighter safety and customer service journey is crew resource management (CRM) applied to the incident management system.

WHAT IS CRM?

CRM is not a new command system, so relax, and don't get nervous. With more than 20 years of use in the airline industry, CRM has paid its dues. If you are looking for a way to eliminate human error from your operations, do read on and pay attention because CRM is just what the doctor ordered.

On December 28, 1978, a commercial airliner needlessly and tragically crashed in Portland, OR. Essentially, the captain of the aircraft did not heed the cautions and warnings expressed by the experienced members of his flight crew. They told the captain that the aircraft did not have enough fuel to compensate for a problem with the landing gear.

The case study that I cited here was the crash of United Airlines Flight 173. The plane ran out of fuel six miles from the airport, resulting in 10 fatalities and 23 injuries. As the emotional stress and pressure [increased] in the cockpit of that DC-8 aircraft, several critical decisions were made that turned out to be disastrous. The CRM program was developed in response to and as a direct result of the collection of human errors that happened aboard United 173 that fateful day.

The Federal Aviation Administration (FAA) led the way to fully develop and implement the comprehensive crew resource management system in response to the last critical element, human error, in preventing commercial aviation crashes. When this human factors performance enhancement system was offered up in 1979, it was called Cockpit Resource Management. After 23 years of use, six major revisions and a name change, "Crew Resource Management" is one of the best human factors engineering and performance enhancement programs available.

CRM COMPONENTS

By way of a brief review of the CRM program, the foundation principle is that to err is human. A formal system of critical checks and balances must be placed into motion to eliminate, or at least reduce the likelihood of, human error. CRM recognizes that humans are prone to making errors and compensates for this fact by adding a second set of eyes, ears and additional brainpower. The design allows for redundancy on the flight line (cockpit) using a challenge-and-confirm mentality to get the decision right the first time every time. It just makes good sense that two heads are better than one when it comes to making critical decisions.

Four tenets/guiding principles are used to frame the CRM concepts:

1. Communications
2. Teamwork/leadership
3. Task allocation
4. Critical decision making

CRITICAL DECISION MAKING

Perhaps the best place to open a discussion about the critical decision making would be to mention the three decision outcome avenues that CRM identifies. The best accident that we can experience is, of course, the one that we can avoid! The philosophy is that with the correct education, training, experience and attitude applied at the proper time, accidents can be avoided before a negative consequence occurs.

The next plan is to use a CRM component to "trap" a bad decision long before there is a price to be paid (accident or injury) by the troops. This second-level outcome is just as good as the first one, if it happens in the appropriate time frame (before an event can occur). Most likely, someone else will be the person trapping another's mistake. Think of this part of the CRM process as having Cal Ripken playing shortstop in the 1980s and 1990s while you are pitching. Nothing gets through to the outfield. The fact is that someone else will likely be your saving grace (catch your error, fig. 10-3), rather than yourself. You would not make the miscue in the first place if you were aware that the consequences of the decision would be an error of any type.

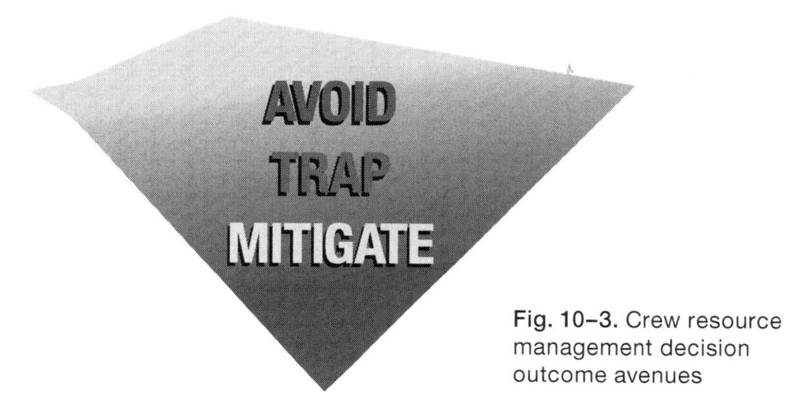

Fig. 10–3. Crew resource management decision outcome avenues

The last CRM priority is to mitigate the results of the effects of an accident or injury after the fact. An analogy is that of milk spilled out of the bottle, and now the only avenue left to fix the situation is to get the roll of paper towels and soak up the damage.

Unfortunately, poor operational decisions have a much greater impact than the need for a few paper towels. According to the type and severity of error, for instance, if an improper tactic or strategy is implemented, the impact could be fatal to our members or customers. This is the least desirable of the matrix levels, but it helps keep the incident commander focused that he or she must be prepared to deal with the results of bad decisions. Actions such as the deployment of the rapid intervention crew, calling additional alarms and clearing tactical radio frequencies for emergency traffic all to come to mind when taking steps to mitigate the impact of a bad decision when losing track of a firefighting crew.

PREPARATION IS THE KEY

The preparation of the command team is perhaps the very best way to avoid an accident from occurring. Being capable of doing the job will increase the ability to make good critical decisions as well.

One area that the commercial airline industry clearly beats us hands down is in the arena of training and certification. Earning that fourth stripe on the sleeve of the pilot's dark-blue blazer takes an untold amount of training and flying experience hours. In fact, on-line pilots are tested in two different ways every year. They have to pass a comprehensive simulator test and then complete a flight satisfactorily under the watchful eye of a FAA evaluator during an unannounced check ride.

Our business must find a way to implement a similar program for incident commanders and their team members to obtain certification to perform these critical job functions. The manner in which paramedics train and recertify is a perfect template that should be considered to provide this command team certification program for the command team. Airline pilot trainers point out that you have to be able to fly the plane before you can make quality critical decisions regarding flight systems (duh!).

The same logic can—must—be used for the preparation of the members of the incident command team. Nationally standardized training and testing standards must be developed for these roles. The National Fire Service Incident Management System Consortium, under the leadership of Seattle Fire Chief Gary Morris, has started an effort to fill this need. This group's purpose is to identify the minimum knowledge, skills, and abilities that one should have before one should be allowed to command an incident of any type. Thank goodness that the airline industry has such standards in place or we would have a lot more emergency disaster work to handle.

Communications preparation must be discussed in any debate about improving critical decision-making. Under ideal conditions, communications between two humans are less than perfect. When my oldest child was preparing to graduate from high school in 1991, he asked for a car upon successful completion of his secondary education. I had the resources and he was a good student both in the classroom and on the athletic field (his major was either girls or sports, I don't remember back that far). So, I agreed to make the purchase to have the vehicle ready for him by his mid-June graduation date. When I said yes, my namesake visualized a new Corvette or perhaps a Porsche 914. I immediately envisioned the 1978 VW Bug being rehabilitated for several hundred dollars. We agreed somewhere in the middle and obtained a pickup truck.

The purpose of this story is to point out that complete and effective communications between people are difficult under ideal conditions. Now add the robust confusion of the incident scene and the framework has been laid for extremely poor communications to exist. Members of the command team (and all fire officers for that matter) must work at and be skilled in effective emergency incident communications.

Finally, under the heading of preparation, we must discuss the need for proper education, training, experience, and attitude. Each one of these elements must be considered in order to be an effective command team member. These personal traits speak beyond the necessary process of required national certification. To steal a Tom Peters axiom, we must "stick to the knitting."

To interpret this truism, we must be able to flawlessly execute the basics of our profession (do the task correctly the first time). This action (great basic performance) enables us to handle response work to meet our customer's needs while keeping our members safe.

STRATEGIC COMMAND POSITIONS

Members of the command team (except the safety officer) must be in the proper location to communicate and to make the critical decisions that will determine the success or failure of the operation. Both the lives of our members and those of our customers weigh in the balance of the incident action plan (or lack of a plan) that command puts into motion.

It is imperative that the command team members set up their operation at a reasonable command post facility and operate using radio headsets tied into their mobile radio units, no exceptions. The use of the mobile radio as compared to the portable radio will improve the engineering component of communications by about sixfold. Consider that the portable operates on about three watts of power, whereas the mobile pushes about 35 watts, meaning that there is no contest of superior electromagnetic wave performance.

When the incident commander carries the axe down the hallway or chops a hole in the roof, it's fun and stress relieving for the field boss, but he or she is totally out of control! No command post (at incidents that exceed one unit) is equivalent to ineffective and inefficient command, placing life and property at greater-than-acceptable risk.

If members of the command team can't operate from an effective command post position, replace them and do it sooner rather than later. Perhaps they should go back to being company officers, or an assignment of all day work attached to the light wagon would be in order. Such behavior (hands-on fire attacking incident commander) will only negatively affect incident operations and everyone's health.

Allowing an "aerobic incident commander" to operate by wandering around at a call would be like letting the air traffic controller operate on the tarmac of the airport's runway. When the pilot needs to depend on the air traffic controller he or she can be found (all of the time for that matter) in the appropriate location to provide the critical help that the pilot relies upon to control the aircraft. That location, of course, is the air traffic control tower that is designed to perform the necessary functions of the job, no exceptions.

Visualize a stormy night and the senior controller announces that he is putting on a raincoat, grabbing a clipboard and a flashlight, and heading to the taxiway to guide a troubled airliner home. The FAA would not punish the deranged controller, but would get him some psychological help for such bizarre behavior. Maybe the "aerobic incident commanders" need a mental health evaluation to ensure that they are fit for duty, after they promise to never repeat the negative performance.

CRM adds a new command post term and concept: the sterile cockpit rule. The sterile cockpit rule means that all unnecessary distractions are eliminated from the work environment to help critical decisions to be made properly. Thirty minutes before take-off, 30 minutes before landing, and anytime an unplanned event occurs, the flying pilot puts sterile cockpit rules into place. No unnecessary or unrelated communications (personal chatting and the like) are allowed to take place during the time that this rule is in effect.

The needed transition is for us to implement and use the sterile cockpit rule at our command post. When we have members that are operating in an immediately dangerous to life and health (IDLH) environment, the incident commander should declare that the sterile command post rule is in effect. Maybe there should be an obvious external indicator (perhaps a flashing blue strobe light) announcing the fact that the sterile command post rule is in effect. Whenever members are in IDLH atmospheres, the command post should operate under sterile command post rules.

SUPPORT OFFICERS' DECISION-MAKING ROLE

Safety officer. The safety officer (SO) becomes the remote "eyes and ears" of the commander and is a principal player of the command team. The SO must get a reasonable view of the risk, if possible. The SO watches out for unsafe acts, unsafe conditions and unsafe behaviors with the goal of correcting any out of balance actions before they can become a negative consequence.

Perhaps this position is the second most important role that plays out at an incident and represents a tremendous informational resource to form critical decisions from during emergencies. It is well established that the SO can shut down or alter operations that indicate imminent accident or injury. Understanding this fact, it is obvious that the person who fills this specific role needs to be educated, trained, experienced and hopefully certified as an incident safety officer. (National certification is available through the Fire Department Safety Officers Association.)

An incident that places members in IDLH environments must have at least one safety officer identified and operating at the scene. The most effective and efficient way to staff this required command team position is to assign a company officer to fill the role of initial safety officer. After visiting many departments, the national trend is to have a single safety officer who is trained and qualified to fill the role. Perhaps the department is enlightened and has a shift safety officer who responds to all emergencies to handle this critical role.

In either situation, often there is a delay in the arrival of that single individual. Therefore, the most efficient and effective way to address this mission critical need is to assign a specific, early-arriving person to the job of safety officer. For instance, a standard operating procedure might identify that the second-arriving engine company officer shall immediately assume the role of safety officer at incidents that involve IDLH atmospheres. The initial SO maintains this responsibility until a more qualified person arrives on location and formally relieves the initial safety officer, at which time he or she can return to the supervision of their company.

The duties and responsibilities of the incident safety officer should be clearly defined, well documented, and trained upon by all members of the department. Without the accessibility to and support from an incident safety officer, high-quality critical decisions will not be able to be made. The incident commander's blind spot (barrier holes) will be too large of a deficit to be able to successfully overcome, resulting in an injury rate that is not acceptable to anyone that serves in fire-rescue operations.

Deputy incident commander. Another required command team position is that of deputy incident commander. The parallel to this job title is that of copilot or first officer. The deputy incident commander adds a clear-cut second-in-command person who possesses all of the knowledge, skills, and abilities of the incident commander. The arrival of this person adds the redundancy element to the command post operations that is necessary to help catch human errors before they have a chance to become a consequence. The deputy incident commander is in a great position to use the challenge-and-confirm process to eliminate errors before they occur.

What a great way to do business when lives hang in the balance of emergency incident decisions. Many fire-rescue agencies use a minimum of two paramedics to deliver advanced life support care. The basic concept is that the risk of making an avoidable and predictable mistake is too high for the agency to assume, so a second paramedic is added to the response resources to provide instant quality assurance. This quality assurance/quality control element has been added to this sophisticated care to lower the risk of mistreating patients in their hour of grave need. The same philosophy exists within the CRM program to help drive out human error. The two-paramedic concept is a textbook example of the utilization of CRM "redundancy on the fight line" concept.

Accountability/documentation officer. Another member of the command team is the accountability/documentation officer. This person must be able to track all members all of the time whenever they are in an IDLH environment. The accountability officer must be able to quickly and correctly answer four questions about the members operating at the incident. The questions are:

1. Who are the members (by name)?
2. Where are the members located (i.e., Division 1)?
3. What actions are they taking (i.e., fire attack)?
4. What conditions are they encountering (i.e., fully involved attack space for 15 minutes)?

With these four vital pieces of information, the incident commander can make higher-quality critical decisions than without this information. Also, by tracking the who, where, and what conditions, it gets tough for members and companies to "freelance" with so many eyes upon them. The best accountability system must have a qualified human to effectively operate it or it has no value at the incident.

The other duty of this position is to be the "scribe" for the incident commander during the battle. The accountability/documentation officer must be able to use check sheets, command boards and any other command enhancement tools found at command posts. The incident commander needs to focus on the ongoing firefight (action plan) and not try to figure out how to put paper in the fax or run the accountability board. This job title is a great justification to add chief's drivers to your system (or keep these positions if you already have them). The duties that are described here are not optional and cannot be performed by one individual.

IN SUMMARY

Crew resource management is a very useful collection of tools for departments to add into their incident command system program. By reviewing the NIOSH firefighter death reports, the need to correct human errors that lead to fatal accidents is painfully obvious. CRM is not a new command system, but in fact, is a specific way that command team members can eliminate human error from ever occurring. The program consists of four major elements: communications, teamwork/leadership, task allocation and critical decision-making.

CRM acknowledges that human errors will happen and cannot be avoided by a single individual. Therefore, a crew approach is best used to manage the resources at hand to resolve all emergencies. A minimum command team (according to your author) should be four members to include the incident commander, deputy incident commander, safety officer, and accountability/documentation officer. The full command team needs to be up and operational at all calls involving IDLH atmospheres.

11

IF YOU DON'T CARE, GET OUT

We may have found a cure for most evils, but we have found no remedy for the worst of them all—the apathy of human beings.

—Helen Keller

One of the most difficult personnel issues to deal with is how to stop members from falling into the "apathy trap," where they decide they don't care about the agency any longer. When a member acts like they don't care about the department or the work that we do, through harmful behavior to other members or, heaven forbid, to customers, a lot of organizational damage can occur. Apathy becomes a slippery slope that can damage both the organization and the member. To not address the apathy can even be more devastating and difficult to manage.

TIPS TO HELP YOU AVOID FALLING INTO THE APATHY TRAP

This section reflects a sincere desire and need to keep your folks engaged and productive throughout their career or affiliation with the department. It is amazing to see how infectious a positive attitude toward the community and our service can be. In direct contrast is how harmful lackadaisical and uninterested behavior can become for all.

I like to ask the most disgruntled employees to take a journey back in time to when they were appointed to the position of firefighter or voted in as a volunteer member. If these people are willing to step back with me, they usually share vivid memories of great community and departmental respect and support. Near the end of the interview process, the disconnected members often make statements about their willingness to take on assignments whenever they are asked, no matter the task involved. Most will recall that they got into this business because they wanted to help the community.

In fact, let's go back in your career for a moment. What was it like when you first realized you were going to be a protector of the people? My guess is that it was a daunting yet exciting feeling. Most folks would remember that time and what it felt like to ride on the big red truck for the first time. I remember the date, time, and address of my first response: it was September 9, 1968. The call came in at 8:30 p.m. for smoke in the building at the Prince George's Plaza in Hyattsville (MD). I was riding on the back step of Engine 12. For me, this was the end of an eight-year wait to be a riding volunteer member, so I am sure you can imagine my excitement and enthusiasm.

Now let's move back to today. What is the "condition of your condition" (your organizational attitude and core value system)? I hope it is the same as it was on your first day in recruit school or when you responded to your first alarm. However, if it is not, what will it take to get you back to that place? Who has control over your morale? Who determines the level at which you participate and contribute within your organization? Who is directly responsible for your personal behavior on and off the job? Who can instill a sense of departmental pride and support? Who makes sure you are self-disciplined and that you stay out of trouble? Who ensures that you are completely capable of performing the job well and that you maintain all of your certifications? Finally, who ensures that you are fit for duty?

Of course, the answer to all of those questions is you. Others can try to influence your actions, decisions, and behaviors, but you are in control of your destiny in your agency and in your life. You have the free will to choose your course and project the attitude and actions that you desire. In fact, the more professional and self-disciplined that you are within your department during stressful times, the more likely you are to be insulated from the potential effects of the negative forces that enter everyone's life from time to time.

Most folks want to be around people who project a positive attitude and image, rather than deal with those who have a lot of baggage and attempt to bring those around them and the organization down to their level. Don't let negative energy get a foothold on you; it will be tough to shake it off. There are many ways to deal with a bad day or even a bad week or year. Perhaps take a vacation, even a small one, to bring about a change of attitude. Communicate your concerns with someone—your supervisor, personal mentor, spouse, or counselor or spiritual advisor—to get a change of perspective and an adjustment of attitude. Try to think about the importance and value of your role within the department and the mission-critical work you do for your community.

FIND A MENTOR

I have had many opportunities to be an adult learner and experience training under many fire service mentors. There are two stories in my repertoire of methods that are crazy good at bringing apathetic members back into the fold of being happy, balanced, and productive folks. In the first story, the chief would spend a minute or two with a disconnected member. During the verbal exchange, this fire chief would give the lethargic and apathetic firefighter a new member application to the fire department.

Often the active (but unproductive) member would express confusion at the suggestion of the department: "Hey Chief, I have worked for you for the past umpteen years!" The chief would then ask the person to rejoin the department and become a fully involved member once again, insisting that the member keep the application form as a reminder—maybe taped to their locker door. Nondestructive, and what a very powerful message the chief of this department was sending with this action.

Another mentor would try to shock or frighten a person back into contributing to the organization. The chief would threaten that folks with bad attitudes who had become disenchanted with the work of the department could go to work at a company he secretly operated called the Fire Department Roofing Company, reroofing and repairing roofs at homes in the community.

The story was told that really poor departmental performers would be banished/assigned to the FD Roofing Company for one week in the middle of the sun's blazing summer. Someone way out of line would be assigned two weeks with the nail apron. If you have ever performed manual labor outdoors in the summer, you know this is far from an enjoyable task. Of course, this company was entirely made up, but the point was still clear: behave, perform, and produce for your department or else.

SUMMARY AND REVIEW

From time to time, a member may need to take a mental or physical break from the work we do. That is expected, normal, very appropriate, and in everyone's best interest. However, once a member enters into a long-running funk and becomes unproductive or apathetic, the leadership must take action.

Consider Abraham Maslow's Hierarchy of Needs and the five levels of motivation he offers. This is a very interesting theory with a large following (including your author) that believes Maslow was spot-on with his work. A quick review of the five steps in the hierarchy: the foundation starts off with the basic physiological needs (food to eat, water to drink, shelter, and all of the basic needs of life). Once the physiological needs are met, the next level moves to safety and security requirements (long-term health, welfare, and survival of the individual). The third level up the chart addresses belonging (being a part of the team, having intimate relationship and friends). The fourth level up the chart is ego or esteem needs (outward signs of personal success, trappings of accomplishment such as badges, collar insignias, trophies). The highest level of attainment is self-actualization and self-fulfillment (achievement of full potential). One of the provisions of this theory is that once a need is met, that specific need is no longer motivational, and the individual reaches for the next level on Maslow's pyramid (fig. 11–1).

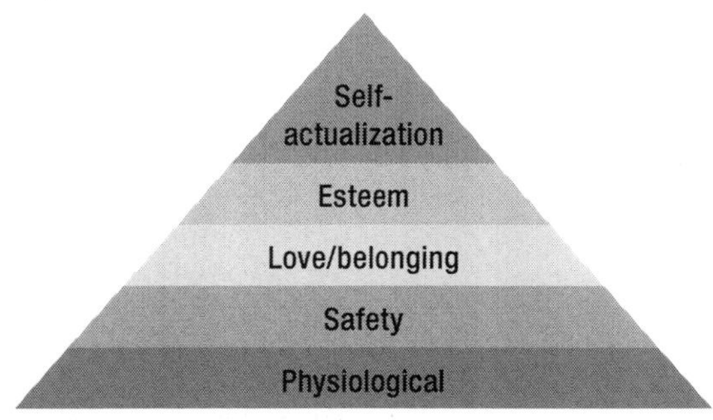

Fig. 1–11. Maslow's Hierarchy of Needs

CRITICAL LEARNING POINTS

1. Lack of caring about the organization by an individual member can become a very difficult issue to resolve.
2. True fire-rescue leaders work to keep members under their command engaged, productive, and positive throughout their careers.
3. Each of us is responsible for our own attitudes. We can choose to be fully involved with our department or not.
4. Consider the various avenues that help a person recover from or avoid a poor attitude about the department: take a vacation, open communications, have a mentor, talk through the issues with someone you trust, transfer to another company and/or shift.
5. Perhaps get your members motivated and rejuvenated: use something similar to the parable of the Fire Department Roofing Company.

LEADERSHIP DISCUSSION QUESTIONS

1. Have you ever been less than fully productive within your department? If so, what was the cause? What snapped you out of this situation?
2. Are there members within your agency who are no longer connected to the work of the department? What actions can be taken to bring about productive change without damaging their career or membership?
3. What is your opinion of the chief's tactic to ask a disconnected member to rejoin the department? How about the threat to assign apathetic firefighters to work for a roofing company as a refresher about the goodness of their job?
4. What are four general ways that leaders can help members through tough times in their lives?
5. Who is responsible for your attitude, behavior, and performance? What does this mean?

PERSONAL LEADERSHIP PLAN FOR IMPROVEMENT

1. What behaviors do you demonstrate that are blocking you from caring about your work performance?

2. What behaviors must you adopt to care about your department?

3. List at least one training program relating to this topic that you will attend in the next year.

4. List at least one leadership textbook relating to this topic that you will read in the next year.

5. List the names of the members of your department you will mentor on the skill of caring.

____

____

____

____

____

12

Manage Your Personal Behavior

Self-discipline is the best discipline for a fire service member or leader.

Organizational discipline is a topic that most folks don't want to discuss or deal with as either a supervisor or subordinate. There is, however, a great deal of interest in and thirst for information that will help the leaders within an agency be consistent, fair, transparent, and honest when required to activate the disciplinary process. This chapter will provide ideas to help everyone make good decisions in regard to their personal behavior.

Whenever I discuss the privileges and pitfalls of taking on the mantle of leadership, the disciplinary process is always a topic of great interest and discussion. Most folks say they are very uncomfortable sitting in judgment, as it were, of other members, but understand the organizational need to maintain discipline. The discussion usually goes something like this: "Being a company (or battalion or division) commander is the best job I've ever had, except for the fact that I am expected to instill and manage the disciplinary process. I work hard to keep myself out of trouble and my nose clean, but I have to investigate situations and deliver punishment when it is needed. A large part of my working day is spent on discipline issues instead of real work. If it weren't for this aspect of my position, I could get at least another meaningful drill in each work week. I dislike this part of my job and wish someone else could handle it for me." Taking the implementation of the disciplinary process to its far reaches, I have had some discussions where a senior officer elected to retire versus being the person in the organization saddled with the management of the discipline process. My only response to these brave men and women is "Welcome to the club." Some of the work that leadership involves isn't as pleasant as the emergency response part.

Let me say loud and clear that the best discipline I know of is *self*-discipline. It seems to me that the best fire-rescue officers and members are the ones who have self-control and self-discipline, on and off the job. Members have the opportunity to learn the rules and regulations of the department as part of their recruit training and orientation process. In fact, the important rules need to be reviewed on the first day on the job. Many organizations have packaged the dozen or so critical rules into one document. This information helps get you through your career. Once you've reviewed the rules, the agency expects you to follow them.

The best way to minimize the amount of discipline you have to hand out is to hire or vote in only people of quality. Idiots, criminals, and military misfits will always cause discipline issues, so don't hire them or recruit them into your department in the first place. Fire and EMS agencies must start with the best possible employee or member attainable. If you are hiring folks with sordid or questionable backgrounds, your agency will likely spend a disproportionate amount of time and energy delivering discipline and possibly going to court to defend your decision. We are in a great position, with so many people interested in becoming firefighters or firefighter-EMTs; you can afford to hire or recruit only the very best your community has to offer. Don't short-change your selection and qualification process; make it the most robust selection system you can afford! This is the period to avoid the headache and heartache of having problem people enter your workforce or membership. If a person was fired from another job or dishonorably discharged (to include less-than-honorable discharge) from the military, it is likely that person will not fit into your outfit. The best quotation I've ever heard on this topic is from a nationally recognized lecturer, retired commander Gordon Graham of the California Highway Patrol: "Past performance will predict future behavior." Pick the best and brightest candidates for public safety positions and never settle for questionable people. Your life will be a lot easier and your department a lot better for the effort.

When a firefighter responds to a medical emergency that requires injuries to be exposed to determine their extent and to deliver treatment, they are expected to cut clothing away from the person's body. When a person dials 911 and requests assistance, firefighters can enter the person's home without a search warrant. Many times, sick or injured people are home alone and their possessions are in plain sight. They believe that the firefighter cutting away the clothing or assisting them while they're sick at home will not violate their trust. All of us, career and volunteer members alike, must possess a very high level of public trust to perform our job duties. Relating to behavior, understand that we are never off duty. I know it may not seem appropriate, or even fair, that fire departments must be concerned about a member's off-duty behavior, but the truth is, some of our members get into a lot of trouble away from the firehouse. If a firefighter robs a bank, beats their spouse, threatens workplace violence, snags a DUI charge, or misuses a weapon, it is difficult to get the public to place all of the necessary trust in the department. Off-duty behavior is just as important as on-duty conduct if we are to maintain our value to those we are sworn to protect.

After a negative event occurs involving our members, the newspaper headline or television news tease starts with the fact that "a firefighter from XY Department" was connected to the crime. It is also generally pointed out if the person is a previous member or a retired member. It is easy to take exception to the way the media covers stories involving fire and EMS members, but the reality is that a higher standard of character is expected of public officials who hold positions of trust in our society. Knowing this, we must meet that challenge or suffer the consequences of our actions. Most departments have structured discipline processes that outline how and when they should be implemented. Once the negative behavior is observed, a formal process starts and needs to be carried out fairly and honestly every time for everybody in the system, no exceptions. The expressed goal should be for the system to correct these behaviors and do the least amount of damage to the member and the department. This is always a balancing act and adds a significant measure of stress to the folks who must implement the disciplinary system.

THE HOT STOVE RULE

The "hot stove rules" come to mind to best explain the disciplinary process in the simplest of terms. Think back to your youth—remember coming home from school when your mother was baking something good for you to eat? Your mother probably delivered some very simple rules for you to live by without being burned by the hot stove she was using to prepare your treats.

The first rule is that the policy should be very clear. In the stove comparison, your mother told you in no uncertain terms that the stove was hot and not to touch it. The department should use written standard operating procedures, standard operating guidelines, and other rules that allow us to do our jobs properly and safely. There also needs to be an appropriate level of training and understanding about the rules that are in place. This must be initial and on-going as it is related to following the rules of the department. This is one area where most outfits fall down. Most will conduct initial training and forget to have continuing education on some of the most important topics of the day: their own policies. I have known agencies that pull out the SOPs and SOGs only to investigate a potential violation. This is not a good practice for the department and not a good way to prepare your people for success.

Returning to the stove analogy, the stove doesn't care about your age, rank, or how great a person you are. If you touch it, it will burn you, regardless of the people you know. The rules, regulations, and policies must be implemented fairly and impartially to all. That means everyone plays by the same rules all of the time; the chief cannot give out passes to folks because the chief likes them or because they will stop being active if they are held accountable.

The effects of touching the hot stove are also immediate: you know you've been burned in less than a second. The discipline systems, particularly in large departments, can take a while to work. The factors to consider, such as a fair and impartial investigation and due process and the rights of the member, are critically important. However, when the process lingers for many months to years, it takes the focus off the discipline and raises the question of whether the process was fair. Discipline should be swift, within reason.

The hot stove always ensures that the punishment fits the crime. If a person merely makes glancing contact for less than a second, the burn is minor. If a person sits on the hot stovetop, however, the burn injury is likely to be severe. The department needs to always strike a balance and ensure that the punishment fits the crime. If the punishment is too lenient, the penalty does not have the desired corrective effect. If the punishment is too much, the system will be viewed as unfair and members will become disheartened and fearful.

Finally, with the hot stove, once the touch is made, the punishment delivered, and the situation is over. The stove does not have a memory that builds a bias against the person who touched it; it will not come after them and burn them without them touching it. This is a difficult part of the human condition: we tend to hang on to issues for a long time. Once the issue is resolved, the supervisor and the department need to move on with a hope that the behavior has been effectively changed.

SUMMARY AND REVIEW

To recap, here are the six hot stove rules to remember and use. This is a good process to measure your system against.

1. Provide members with clear policies and proper training.
2. Be impartial; do not play favorites.
3. Discipline errant members immediately.
4. Make sure the punishment fits the crime.
5. Let the discipline be the end of the situation.

Do your department and yourself a favor and select the best folks to become members (never just settle). Next, try to instill self-discipline at all levels of your department. Finally, have a solid discipline system that is fair, open, transparent, consistent, and honest.

CRITICAL LEARNING POINTS

1. Self-discipline will always be the best discipline in every circumstance.
2. Hire or vote in good people of all types. Never hire idiots, criminals, or military misfits.
3. When discipline is necessary, it must be fair, honest, transparent, and consistent for all.
4. Use external discipline as a last resort.
5. The focus should be on doing the least harm to the member and the most good for the department.
6. Remember the hot stove rules.

LEADERSHIP DISCUSSION QUESTIONS

1. Describe an organization that has an unfair discipline process. Please provide the details and background without using the specifics (no names, dates, or places, please).
2. Describe an organization that has a fair discipline process. Please provide the details and background without using the specifics (no names, dates, or places, please).
3. Describe the perceived differences between these two disciplinary processes (fair and unfair).
4. Several times I mention not to hire idiots, criminals, or military misfits. Do you agree with this advice? Why or why not?
5. Why is it important for the department to possess and maintain the public trust?
6. How does organizational discipline help or hinder maintaining the public trust?

PERSONAL LEADERSHIP PLAN FOR IMPROVEMENT

1. What behaviors do you demonstrate that are blocking you from managing your personal behaviors for all your position requirements?

____

____

____

____

2. What behaviors must you adopt to manage your personal behaviors within your department?

____

____

____

____

3. List at least one training program relating to this topic that you will attend in the next year to become a better leader for your department.

____
____
____

4. List at least one leadership textbook relating to this topic that you will read in the next year to become a better leader for your department.

____
____
____

5. List the names of the members of your department that you will mentor on the skill of managing personal behaviors.

____
____
____
____

13

Be Nice

I have also decided to stick with love, for I know that love is ultimately the only answer to mankind's problems.... Hate is too great a burden to bear.

—Dr. Martin Luther King Jr., Pastor and Civil Rights Leader

It is interesting that some aspects of the fire and EMS world are constantly becoming a bit more complicated. When I started in our business as a career recruit firefighter-EMT, fire station work was fairly simple. Much of the life-saving technology we take for granted today did not exist. But some things have not changed, and this chapter will explore one of those pillars of great customer service: being nice. This is a principle that we do not want to ever change.

The concept of being nice leads back to the writings and teachings (once again) of Chief Alan Brunacini. Chief Bruno can go on for hours about the impacts and effects that simply being nice has on our customers, on our departments, and on us. No leadership book on personal rules of behavior and conduct would be complete without making mention of this golden rule. Years ago, Chief B passed out small reflective stickers that urged the reader to "Be Nice." I stuck one on my suitcase by the handle and in the center of the lid as well. I bet that I have had hundreds of folks comment on how cool the chief's message is. In fact, I have asked for some extra stickers and have distributed more than I can count.

The goal of every fire-rescue department is to be a high-performance and high-trust agency known for its fairness, openness, and transparency, where good personnel morale is important to all. Being nice takes on a very significant role in reaching these benchmarks. Great organizations have both the minds and hearts of the members and customers. This core value is the basis or foundation of reaching the lofty goals described above.

Think of the meanest kid in your neighborhood. Now, go on a virtual trip with me to the mean kid's house and knock on their door. Who is most likely to open the door? My sense is that it will be a mean mom or dad or perhaps a mean older sibling. The point I want to make is that the mean kid had to learn how to be mean from someone and that it is typically mom, dad, or older brothers and sisters.

The fire chief expects high quality and trustworthy performance from all department personnel, all of the time, and under all conditions. Therefore, the chief must be nice to the members who answer the calls all hours of the day and night and under all conditions. If the chief treats their people in a mean way or if the organization is not supportive of the members, it is a real stretch of the imagination to think that these firefighters will go out and be nice. The outfit is most likely going to receive complaints about poor performance and the negative attitude displayed by the workforce. However, if the department supports and treats the workforce nicely, it's more likely the firefighters will meet or exceed the challenges that you put before them in the most professional way.

Remind your members that they must be nice first. The organizational expectation is that your folks will be nice to other members and customers as the interactions are initiated, without waiting for the other party to be nice first. It is amazing how many negative issues can be avoided if the interaction starts on a positive footing. You must remember that your customers are generally having a really bad day when they dial 9-1-1, and they typically have very little experience with emergency events. On the other hand, resolving emergencies and not adding stress to the customer's day is our vocation as soon as we pin on the badge.

Your members should feel empowered to be nice to one another and to your customers. In organizations that are well supported, the list of success stories about acts of kindness is just about endless, with tales of matching departmental support from budget dollars to private donations to the old standby, a simple thank you. Several fire departments produce a quarterly "great news" book: a collection of the thank-you notes and cards received during the previous 90 days. This publication is shipped to the mayor, city council, and local media, giving these external stakeholders some idea of what is happening inside the organization. The great news book is added to the department's website so that their membership and others can quickly access the information.

Some departments have a customer service award process that recognizes members for going above and beyond. These types of programs should never interfere with the various types of recognition programs that are in place for fire, medical, and rescue personnel; rather, being nice should be an embellishment added to a departmental awards program.

The first part of high-quality customer service is being able to perform your job flawlessly (Rule 3). Nothing can replace being a high-performance emergency medical technician, paramedic, and/or firefighter. Know your job inside and out and do it well, every time. Being nice will be a major added bonus for you, your department, and the citizens and visitors who call on you.

I once learned that one of our ladder companies had responded to a water leak on the second floor of a low-income apartment project. No surprise there. (If you remember the story about "deer roast with a smile," this involves the same company commander.) As the story goes, the ladder truck was dispatched to a water leak: water coming through the ceiling into the kitchen of a first-floor apartment unit. Upon arrival the ladder company was confronted with a large amount of water leaking onto the floor, mostly dripping from the light fixture. The captain had his crew start the clean-up process in the unit while he checked on the unit directly above to see what was causing the problem.

The captain gave a quick knock on the door and it was pulled opened by a lady, with two small children looking on. The fire captain asked the occupant if she was experiencing a water leak in her apartment. The lady responded that the water hose to her washer had split open and that she was doing the very best that she could to prevent the leak, but it was overwhelming. The captain asked to see the leaking hose, and sure enough, the hose was dry-rotted and split open behind the coupling. Without hesitation, the captain quickly reached behind the malfunctioning appliance and closed the hand wheel to stop the leak.

The young mother of two little ones pointed out that she needed clean clothes to go to work in and for her children to go to daycare. So, the captain and crew did an amazing act of kindness. The ladder truck boss asked the her to leave the water turned off and to make do with her wardrobe for one more day. In return, the captain promised to purchase a new water connection hose for her washing machine and install it for her the next evening (his day off). She agreed. With the water off, apartment downstairs was cleaned up, and the ladder was headed back to quarters soon after.

The captain purchased the replacement washing machine hose and did the quick repair once the customer returned home the next evening. The cost was $6 or $7 bucks (reimbursed by the city out of petty cash). Everyone went home in a Cadillac (all needs were met). The naysayer would ask why should the city (department) take on this level of responsibility. The reality of the matter is that if the captain had not made the repairs, that ladder company would have been called back to that location at all hours of the day and night. It was clear to the captain that this lady was going to do the wash even though the broken hose was leaking. The apartment management was placed on notice about the repair, but in this specific situation (the projects) it would have taken several weeks to get the new one ordered and installed.

This is a clear-cut example of being nice, providing customer service and doing the right thing for the right reasons.

SUMMARY AND REVIEW

"Be nice" is a simple phrase. Easy to understand and most times easy to implement. The results and benefit of being nice cannot be overestimated. We need to be nice to everyone. We need to be nice first, instead of waiting for the other party to be nice. Folks in leadership positions need to become the poster children for the rule. Remember, your followers are watching every move you make.

CRITICAL LEARNING POINTS

1. We need to focus on the technical work (being able to flawlessly do our job—Rule 3) and at the same time treat the customers nicely.
2. Usually, mean kids learn to be mean from mean parents. Therefore, treat your members well, so they are focused on treating the customers well.
3. Being nice first has many advantages. The customer will be glad you started the process (in most cases) and will respond in kind.
4. Consider adopting your own departmental empowerment rules for being nice. The posters were a consistent reminder to do good things for our customers.
5. Remember the case studies described in this chapter and throughout this textbook. It costs very little to add value to the response and to the lives of our customers. Being nice is the right thing to do!

LEADERSHIP DISCUSSION QUESTIONS

1. Describe and detail a situation where you or your department went above and beyond what was expected. How much extra did this service cost you and your department? What was the customer's opinion of your service? How did your fire company feel toward delivering above and beyond?
2. Have you ever received poor customer service? If so, describe it. How did you feel about it? Whom did you tell about it? Ever use social media to express your customer service disappointment?
3. Compare and contrast questions 1 and 2. The theme of your responses is the benefits good customer service and the pitfalls of bad customer service.

PERSONAL LEADERSHIP PLAN FOR IMPROVEMENT

1. What behaviors do you demonstrate that are blocking you from being nice in all your position requirements?

2. What behaviors must you adopt to be nice within your department?

3. List at least one training program relating to this topic that you will attend in the next year to become a better leader for your department.

4. List at least one leadership textbook relating to this topic that you will read in the next year to become a better leader for your department.

5. List the names of the members of your department that you will mentor on the skill of being nice.

REFERENCES

Civil War Trust. 2017. "Biography: Lewis Armistead." CivilWar.org. https://www.civilwar.org/learn/biographies/lewis-armistead.

Covey, Steven R. 1989. *7 Habits of Highly Effective People*. New York: Simon and Schuster.

Hermann, Peter. 2014. "Report on Death of Man Outside D.C. Fire Station: Five Firefighters Should be Disciplined." *Washington Post*, February 21. https://www.washingtonpost.com/local/crime/report-on-death-of-man-outside-fire-station-says-five-firefighters-should-be-disciplined/2014/02/21/f35f111a-9a2a-11e3-b88d-f36c07223d88_story.html?utm_term=.03f06435322e.

NBC News. 2014. "Ferguson Cop Darren Wilson Will Never Police Again, His Lawyer Says." November 27. https://www.nbcnews.com/storyline/michael-brown-shooting/ferguson-cop-darren-wilson-will-never-police-again-his-lawyer-n257511.

New Jersey Bureau of Fire Safety. 1988. *Fire Investigation Report: Firefighter Fatalities; Hackensack Ford.*

Powell, Colin, with Joseph E. Persico. 2003. *My American Journey*. Rev. ed. New York: Ballantine Books.

Quan, Holly. 2011. "Alameda Police Release Memorial Day Drowning 911 Calls." KCBS. http://sanfrancisco.cbslocal.com/2011/06/08/alameda-police-release-memorial-day-drowning-911-calls/.

Rubin, Dennis. 2002. "Crew Resource Management - Part 6: Critical Decision Making." *Firehouse*, July 1. http://www.firehouse.com/article/10545317/crew-resource-management-part-6-critical-decision-making.

———. 2005. "Keeping the Promise: The Fireman's Fund Re-ignited." Firehouse.com, June 1. http://www.firehouse.com/article/10510191/keeping-the-promise-the-firemans-fund-re-ignited.

———. 2006. "Why Now Is the Time for CRM in the Fire Service." Firehouse.com, September 1. http://www.firehouse.com/article/10504425/why-now-is-the-time-for-crm-in-the-fire-service.

———. 2007. "Preventing Harm, One Home at a Time." *Firehouse Magazine*, November. http://www.firehouse.com/article/10503651/preventing-harm-one-home-at-a-time.

———. 2010. *Washington DC Fire and EMS Department: 2011 Transition Plan.* Presented to Mayor Adrian Fenty and the Department Membership, October 7. http://www.fireengineering.com/content/dam/fe/online-articles/documents/2013/2011%20Fire%20and%20EMS%20Department%20Transition%20Plan.pdf.

———. 2013a. "Case Study: Lessons from a Freak-Accident Rescue." FireRescue1.com, February 19. https://www.firerescue1.com/disaster-management/articles/1407368-Case-study-Lessons-from-a-freak-accident-rescue/.

———. 2013b. "How to Gain Complete Compliance for SOGs." Tip of the Spear, FireRecruit.com, May 21. https://www.firerecruit.com/articles/1449919-How-to-gain-complete-compliance-for-SOGs.

———. 2014a. "5 Rules to Being a Great Mentor." Tip of the Spear, FireRescue1.com, January 21. https://www.firerescue1.com/cod-company-officer-development/articles/1653033-5-rules-to-being-a-great-mentor/.

———. 2014b. "Case Study: Fire, EMS Response to Active Shooter." Tip of the Spear, EMS1.com, July 1. https://www.ems1.com/fire-ems/articles/1941660-Case-study-Fire-EMS-response-to-active-shooter/.

———. 2015. "Want to Flame Out Your Fire Service Career, Tell Lies." Tip of the Spear, FireRescue1.com, January 13. https://www.firerescue1.com/cod-company-officer-development/articles/2081707-Want-to-flame-out-your-fire-service-career-tell-lies/.